D1511720

Solar Power

by Peggy J. Parks

Energy and the Environment

ReferencePoint Press®

San Diego, CA

© 2010 ReferencePoint Press, Inc.

For more information, contact:
ReferencePoint Press, Inc.
PO Box 27779
San Diego, CA 92198
www.ReferencePointPress.com

Picture credits:
Cover: IStockphoto.com
Maury Aaseng: 32–35, 47–49, 63–65, 77–79
AP Images: 16
IStockphoto.com: 12, 77 (background)

LIBRARY OF CONGRESS CATALOGING-IN-PUBLICATION DATA

Parks, Peggy J., 1951–
    Solar power / by Peggy J. Parks.
       p. cm. — (Compact research series)
    Includes bibliographical references and index.
    ISBN-13: 978-1-60152-074-6 (hardback)
    ISBN-10: 1-60152-074-3 (hardback)
    1. Solar energy. I. Title.
    TJ810.P38   2009
    333.792'3—dc22

                                            2008053327

Contents

Foreword 4

Solar Power at a Glance 6

Overview 8

Is Solar Power a Viable Energy Source? 20
Primary Source Quotes 28
Facts and Illustrations 31

Could Solar Power Replace Fossil Fuels? 36
Primary Source Quotes 43
Facts and Illustrations 46

Could Solar Power Help Stop Global Warming? 50
Primary Source Quotes 58
Facts and Illustrations 62

What Is the Future of Solar Power? 66
Primary Source Quotes 73
Facts and Illustrations 76

Chronology 80

Key People and Advocacy Groups 82

Related Organizations 83

For Further Research 87

Source Notes 90

List of Illustrations 92

Index 93

About the Author 96

Foreword

"Where is the knowledge we have lost in information?"

—T.S. Eliot, "The Rock."

As modern civilization continues to evolve, its ability to create, store, distribute, and access information expands exponentially. The explosion of information from all media continues to increase at a phenomenal rate. By 2020 some experts predict the worldwide information base will double every 73 days. While access to diverse sources of information and perspectives is paramount to any democratic society, information alone cannot help people gain knowledge and understanding. Information must be organized and presented clearly and succinctly in order to be understood. The challenge in the digital age becomes not the creation of information, but how best to sort, organize, enhance, and present information.

ReferencePoint Press developed the *Compact Research* series with this challenge of the information age in mind. More than any other subject area today, researching current issues can yield vast, diverse, and unqualified information that can be intimidating and overwhelming for even the most advanced and motivated researcher. The *Compact Research* series offers a compact, relevant, intelligent, and conveniently organized collection of information covering a variety of current topics ranging from illegal immigration and deforestation to diseases such as anorexia and meningitis.

The series focuses on three types of information: objective single-author narratives, opinion-based primary source quotations, and facts and statistics. The clearly written objective narratives provide context and reliable background information. Primary source quotes are carefully selected and cited, exposing the reader to differing points of view. And facts and statistics sections aid the reader in evaluating perspectives. Presenting these key types of information creates a richer, more balanced learning experience.

For better understanding and convenience, the series enhances information by organizing it into narrower topics and adding design features that make it easy for a reader to identify desired content. For example, in *Compact Research: Illegal Immigration*, a chapter covering the economic impact of illegal immigration has an objective narrative explaining the various ways the economy is impacted, a balanced section of numerous primary source quotes on the topic, followed by facts and full-color illustrations to encourage evaluation of contrasting perspectives.

The ancient Roman philosopher Lucius Annaeus Seneca wrote, "It is quality rather than quantity that matters." More than just a collection of content, the *Compact Research* series is simply committed to creating, finding, organizing, and presenting the most relevant and appropriate amount of information on a current topic in a user-friendly style that invites, intrigues, and fosters understanding.

Solar Power at a Glance

Solar Power and the Obama Energy Agenda

In February 2009, President Barack Obama signed the American Recovery and Reinvestment Act, a $787 billion economic stimulus bill that includes $1.6 billion in investment tax credits for solar and other alternative energy sources. Additionally, the Obama administration's "New Energy for America" plan sets out long-term energy goals such as ensuring that 25 percent of America's electricity comes from renewable sources by 2025 and investing $150 billion over 10 years to stimulate private clean energy projects. The how, what, and when of such spending has prompted considerable debate.

The Sun as an Energy Source

According to the National Energy Education Development (NEED) Project, the sun radiates more energy in one second than the world has used since the beginning of time.

A Natural Filter

The atmosphere is the single greatest factor in controlling how much sunlight reaches Earth's surface, which prevents the planet from becoming too hot or too cold.

Harnessing Solar Power

The sun's energy is captured in various ways; for example, solar thermal systems capture the sun's heat, and photovoltaic (PV) technology captures the sun's light.

Use of Solar Power

Solar energy can be used to heat spaces such as homes, offices, schools, and other buildings, as well as to heat water. Large-scale concentrating solar power plants and photovoltaic power plants use the sun's energy to generate electricity. Satellites and other spacecraft, including the Hubble Space Telescope and the International Space Station receive their electricity from the sun.

Advantages

Solar power is a renewable energy source, meaning that it will not be used up until the sun stops shining (about 5 billion years from now). It is also clean and emits no pollutants into the air or water.

Viability as Energy Source

Solar power's potential has barely begun to be tapped. Many scientists are convinced that the sun has the capability of providing nearly all of the world's energy needs in the future. Yet there are challenges, such as the high cost of building solar operations, storage, and transmission from desert installations to cities that are far away.

Solar Power Versus Fossil Fuels

Fossil fuels currently provide 80 to 90 percent of the world's energy needs, but they are a finite resource, meaning that they will someday be depleted. A growing number of scientists believe that by transitioning to solar power, the need for fossil fuels will shrink, and this will vastly reduce emissions of heat-trapping gases into the atmosphere.

Solar Power in the Future

Scientists are constantly looking for new ways to utilize energy from the sun, such as continuing to build massive solar power plants throughout the world, increasing the use of solar thermal power, and expanding solar power to vehicles.

Overview

For thousands of years humans have known about the immense potential of the sun, and they were aware that their livelihoods depended on it. In ancient times people learned that they could use magnifying glasses or mirrors to concentrate sunlight into beams that were hot enough to spark kindling used to build fires for warmth and cooking and also to light torches for religious ceremonies. The ancient Romans warmed their famous bathhouses by facing large windows toward the south to let in the sun's heat, and native peoples in North America later did the same with their houses. During the eighteenth century Swiss scientist Horace de Saussure built the world's first solar collector, which he called a "hot box." In writing about his observations, Saussure commented on how the sun's heat was amplified: "It is a known fact, and a fact that has probably been known for a long time, that a room, a carriage, or any other place is hotter when the rays of the sun pass through glass."[1] Expanding on Saussure's design, noted astronomer Sir John Herschel built his own hot box, which he used to cook food during his South African expedition in the 1830s.

Today, with scientists pointing toward the sun as one of the most promising energy sources, the earliest uses of solar power may seem rudimentary and primitive. Yet looking back in time provides a glimpse at how crucial the sun has always been for human life. As technology continues to become more sophisticated, there is every reason to believe that solar power will grow even more important in the future.

A Powerful Energy Source

When the sun is shining brightly in the daytime sky, warming Earth and bathing it in light, it is difficult to imagine that it is just another star. It is, though, and scientists say that when compared with other stars, the sun is actually rather ordinary. But it is unique because even though it is 93 million miles (150 million km) away, the sun is closer to Earth than any other star in the universe; thus, it is often called Earth's star. And even though many stars are larger, the sun is unbelievably massive. Measuring about 865,000 miles (1.4 million km) in diameter, the sun dwarfs Earth, making the planet seem like nothing more than a tiny pebble in comparison.

The energy from the sun is created deep within its core, through nuclear fusion reactions that are caused by immense heat and pressure. In a process known as solar radiation, the sun's rays radiate outward into space, traveling in the form of concentrated particles of energy known as photons, which are smaller than atoms and are invisible to the human eye. As distant as the sun is, it only takes

> " For thousands of years humans have known about the immense potential of the sun, and they were aware that their livelihoods depended on it. "

about eight minutes for sunlight to travel to Earth—but not all of it strikes the surface. By far, the biggest factor that affects how much sunlight reaches Earth is the atmosphere, which filters out a certain amount of solar rays. Other influences include the time of year, the time of day, and the amount of cloud cover, as well as the latitude of a particular area. Countries closest to the equator, for instance, receive more constant, direct sunlight and are therefore hotter than areas farther away.

Untapped Potential

Scientists have long known that the sun holds vast energy potential, but only a fraction of it is currently being utilized. According to the Energy Information Administration (EIA), less than 7 percent of energy consumption in the United States came from renewable energy sources in 2007, and the smallest amount (0.1 percent) was from solar power. This is in stark contrast to fossil fuels, which the EIA says represented 86 percent of total energy consumption during the same year.

Noted solar power authorities Ken Zweibel, James Mason, and Vasilis Fthenakis, say that the sun has vast potential that has barely begun to be tapped, as they explain: "Solar energy's potential is off the chart. The energy in sunlight striking the earth for 40 minutes is equivalent to global energy consumption for a year."[2] The scientists add that the United States is fortunate to have a massive amount of land that is suitable for building solar power plants—at least 250,000 square miles (648,000 square km) of land in the Southwest alone, where immense amounts of solar radiation reach the surface every year. "Converting only 2.5 percent of that radiation into electricity," they write, "would match the nation's total energy consumption in 2006."[3]

How Solar Power Is Captured and Used

Various methods are used to harness the sun's energy, such as solar thermal systems, which use reflective solar collectors to capture and focus the sun's heat. Solar thermal systems can heat water for swimming pools as well as for bathing and washing clothes. In Golden, Colorado, the Jefferson County Jail uses solar thermal energy to heat its water, and the system provides the facility with at least half of its hot water needs. Other uses for solar thermal systems include heating spaces such as homes, offices, schools, and other types of buildings. Large solar thermal power operations, known as concentrating solar power plants, use solar energy to generate electricity. Energy consultant Virginia Lacy explains the science behind this sort of operation: "Remember when you were a kid and you used a magnifying glass to concentrate sunlight on an object to set it on fire? Solar thermal electric plants apply the same principle on a grand scale."[4] Huge mirrors or other reflective devices focus sunlight onto a receiver that heats liquid to high temperatures, which in turn forms steam that turns electrical generators.

Another way of generating electricity is a type of technology known as photovoltaics, which comes from the words *photo*, meaning "light," and *volt*, which is a measurement of electricity. PV systems use solar cells that are made of silicon-based material to absorb the sun's light, rather than heat as with solar thermal systems. As photons in the sunlight are absorbed, electrons are knocked loose from atoms in the silicon, and the resulting flow of electrons through the material generates electricity. One example of photovoltaics is calculators that need no batteries and continue to operate as long as they are exposed to light.

> **According to the Energy Information Administration (EIA), less than 7 percent of energy consumption in the United States came from renewable energy sources in 2007, and the smallest amount (0.1 percent) was from solar power.**

Expanding on this basic technology, large photovoltaic power plants capture sunlight with solar cells and convert it directly into electricity. One of the biggest solar photovoltaic plants in the world, located in Jumilla, Murcia, Spain, provides enough electricity for the equivalent of about 20,000 homes. The largest photovoltaic solar power plant in North America is located at the Nellis Air Force Base near Las Vegas, Nevada. It is a massive operation that is spread over more than 140 acres (56 ha), and it provides 30 percent of the electrical needs on the base where 12,000 people work and more than 7,200 people live.

Nature's Solar Collectors

Earth's atmosphere, oceans, land, and plant life work together to attract, absorb, and distribute solar power. The atmosphere, for example, acts as a protective blanket that plays a crucial role in controlling the amount of

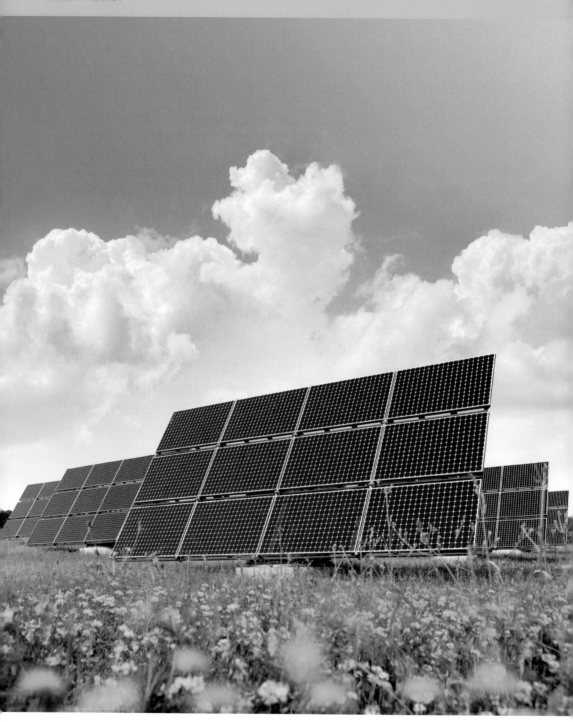

These solar panels are capturing the sun's light to produce energy. Solar power is a renewable energy source, meaning that it will not be used up until the sun stops shining—about 5 billion years from now.

solar energy that reaches the surface. If too much sunlight were able to get through the atmosphere, the planet would burn up, and if too little sunlight reached the surface, Earth would be a frozen wasteland. The oceans, which cover more than 70 percent of the planet, work along with the atmosphere to regulate the climate and make it hospitable for life. Through a chemical reaction known as photosynthesis, plants absorb energy from the sun and carbon dioxide (CO_2) from the atmosphere to create carbohydrates for food, as well as oxygen, and then "exhale" the oxygen back into the air.

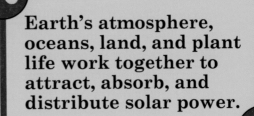

> " Earth's atmosphere, oceans, land, and plant life work together to attract, absorb, and distribute solar power. "

The sun's energy also drives winds that can be captured to generate electricity. Wind turbines, mounted on enormous towers, have three propeller-like blades at the top. As these blades spin they collect the wind's kinetic energy, which is the energy of motion. The spinning blades turn a driveshaft, the driveshaft transfers the kinetic energy to a generator, and through this process electricity is created.

Solar Power in Space

Not only is the sun's energy captured and used on the ground, it also powers spacecraft orbiting high above Earth. The Hubble Space Telescope, an enormous, space-based orbiting observatory, receives its electrical power through systems of solar panels known as arrays that are mounted on the outside. Since it was launched in April 1990, the Hubble has operated around the clock, beaming hundreds of thousands of spectacular space photos back to NASA's Goddard Space Flight Center in Maryland. By studying this data throughout the years, scientists have gained valuable knowledge about the universe.

Solar energy is used to power the electrical systems of satellites, as well as space shuttles once they are in orbit. The International Space Station (ISS), which orbits at an average of 220 miles (354 km) above Earth, also receives its electricity from the sun. Eight massive, powerful solar arrays are mounted on the outside of the ISS, and on each are 2 blankets of solar cells that capture sunlight and convert it into electricity that is needed to power the spacecraft, including its onboard laboratories where astronauts and cosmonauts conduct scientific experiments.

Advantages of Solar Power

Tapping into the sun's energy has numerous positive aspects. Solar power is a renewable resource, meaning that it will continue to be available until the sun burns out—and scientists say that will not happen for about 5 billion years. Solar energy plants consume little or no fuel, which could potentially save tremendous amounts of money that is currently spent to run coal-fired power plants. For instance, the photovoltaic plant at Nellis Air Force Base saves the military an estimated $1 million per year in electricity costs. Another significant benefit of solar power is that it does not harm the environment; rather, it is clean energy that emits no pollutants into the air or water.

In a report titled *US Solar Industry Year in Review 2007*, the Solar Energy Industries Association (SEIA) and the Prometheus Institute laud the economic advantages of using solar power. The report explains: "In many applications today, solar energy in a home or business, when properly installed and financed, can immediately begin to save money on energy bills. . . . For utilities, solar energy can provide valuable intermediate and peak load power. Also, for utilities with an aging transmission and distribution infrastructure, distributed solar can help stabilize grids and offset expensive infrastructure upgrades."[5] The SEIA adds that another benefit is the creation of numerous jobs in solar power industries such as manufacturing and distribution operations, as well as trade jobs for engineers, designers, electricians, plumbers, and roofers.

Is Solar Power a Viable Energy Source?

As promising as solar power is, it is not without its challenges, as photovoltaic specialist Paula Mints explains: "The industry has some problems to solve. Solar energy has been around 30 years and is still a start-up industry."[6] One of the biggest challenges is that large solar power operations are extremely expensive to build. It cost more than $100 million to build the Nellis Air Force Base solar power plant, and one that is under construction near Gila Bend, Arizona, is projected to cost more than $1 billion.

Another challenge posed by solar power is that huge amounts of unobstructed space are required to build large solar energy operations.

Desert areas are ideal locations because the sun shines almost constantly during the daytime. If numerous plants are built in deserts, however, this would take up massive amounts of space and inevitably cause damage to the environment and wildlife. Also, even if these plants are constructed, it is no easy task to transmit electricity from desert areas to cities that are far away. That would require the construction of huge storage units, most likely underground, as well as some sort of direct-current power backbone so that energy could be efficiently sent across the country. Such an undertaking would be extremely costly.

Scientists and energy experts agree that many challenges need to be worked out in order for solar power to become a major energy source. However, many say that technology has already begun to overcome some of the biggest hurdles, and it will continue to do so in the future.

Could Solar Power Replace Fossil Fuels?

Coal, crude oil, and natural gas are known as fossil fuels because they formed over hundreds of millions of years from the fossilized remains of prehistoric organisms. The U.S. Department of Energy (DOE) explains: "When these ancient living things died, they decomposed and became buried under layers and layers of mud, rock, and sand. Eventually, hundreds and sometimes thousands of feet of earth covered them. In some areas, the decomposing materials were covered by ancient seas, then the seas dried up and receded. During the millions of years that passed, the dead plants and animals slowly decomposed into organic materials and formed fossil fuels."[7] Humans began to discover these fuels during ancient times and used them in a variety of ways to greatly improve their standard of life.

> 66
> **Solar power is a renewable resource, meaning that it will continue to be available until the sun burns out— and scientists say that will not happen for about 5 billion years.**
> 99

Today fossil fuels are used for everything from generating electricity at power plants to heating buildings and fueling transportation. Although people all over the world depend on them, these fuels are often a source

Members of the United Nations Framework Convention on Climate Change arrive in a solar-powered taxi. Solar power is clean and emits no pollutants into the air or water.

of controversy because when they are burned, pollutants are released into the air and water. For instance, coal-fired power plants emit sulfur dioxide gas; when the gas is released into the atmosphere, it can create acid rain, which kills vegetation as well as fish and other wildlife. Another toxic gas, nitrogen dioxide, is also emitted by coal-fired power plants, but an even bigger source is automobiles, trucks, buses, trains, and other forms of transportation that emit the pollutants through their exhaust systems.

Yet even if there were no environmental problems caused by fossil fuels, relying on them as an energy source poses another challenge for the future: They are finite resources, meaning that they will someday be used up. In the October 2008 report *Energy Revolution*, lead author Sven Teske writes: "Supplies of all fossil fuels—oil, gas and coal— are becoming scarcer and more expensive to produce. The days of 'cheap oil and gas' are coming to an end. . . . By contrast, the reserves of renewable energy that are technically accessible globally are large enough to provide about six times more power than the world currently consumes— forever."[8] Many scientists agree that renewable energy sources, including solar power, have the potential to vastly reduce the world's dependence on fossil fuels, and perhaps even replace them altogether.

> " Since the mid-eighteenth century, when the Industrial Revolution began in Europe, humans have emitted billions of tons of carbon dioxide (CO_2) into the atmosphere by burning fossil fuels. "

Could Solar Power Help Stop Global Warming?

Global warming is a highly controversial topic. It refers to the increase in the average global temperature, which many scientists are convinced is rising more rapidly than ever before in history. Their concern is that human actions—specifically the burning of fossil fuels—have caused the Earth to warm at an unprecedented rate. Since the mid-eighteenth century, when the Industrial Revolution began in Europe, humans have emitted billions of tons of carbon dioxide into the atmosphere by burning fossil fuels. CO_2 is known as a heat-trapping, or greenhouse, gas, because when high levels remain in the atmosphere, this traps solar heat, which can cause the land and oceans to become warmer than normal. The Union of Concerned Scientists issued a warning about what will happen if this is not stopped: "We're treating our atmosphere like we once did our rivers. We used to dump waste thoughtlessly into our waterways, believing that they were infinite in their capacity to hold rubbish. But when entire fisheries were poisoned and rivers began to catch fire, we realized what a horrible mistake that was. Our atmosphere has limits too. CO_2 remains in the atmosphere for about 100 years. The longer we keep polluting, the longer it will take to recover and the more irreversible damage will be done."[9]

Scientists often disagree about how serious global warming is, how much of a role humans have played in its development, and what it means for the future. Some even doubt that global warming is a problem at all, saying that it is just another phase in natural climate cycles that Earth has undergone throughout history. But those who are most concerned about it say that if solar power were to permanently replace fossil fuels, it would have a huge impact on reducing environmental damage and could help slow down global warming to protect the planet for future generations.

> **Scientists often disagree about how serious global warming is, how much of a role humans have played in its development, and what it means for the future.**

What Is the Future of Solar Power?

The many ways that solar power is already being utilized provides a glimpse of what the future holds. As the National Energy Education and Development (NEED) Project writes: "Solar energy has great potential for the future. Solar energy is free, and its supplies are unlimited. It does not pollute or otherwise damage the environment. It cannot be controlled by any one nation or industry. If we can improve the technology to harness the sun's enormous power, we may never face energy shortages again."[10]

Scientists are constantly exploring and testing new ways that solar power can be harnessed and used. State-of-the-art solar power plants are being constructed all over the world, and many are expected to be operational no later than 2011. New technology is being developed that will power vehicles, including cars and boats. One futuristic idea that has been under consideration for many years is that satellites could collect solar energy in space and beam it back to Earth. If this technology were to be perfected, it would overcome the problem of some areas of the world having sunlight levels that are too low to generate ample solar energy. As scientist John C. Mankins writes: "The sun shines continuously in space. And in space, sunlight carries about 35 percent more energy than sunlight attenuated by the air before it reaches the Earth's surface. No weather, no nighttime, no seasonal changes; space is an obvious place to collect energy for use on Earth."[11]

Because of the challenges still associated with solar power, no one can say for sure what tomorrow holds or whether the sun will ever become the world's number-one energy source. But as fossil fuels continue to be depleted, concerns about the environment grow, and people search for ways to save money on energy costs, solar power holds a great deal of promise for the future.

Is Solar Power a Viable Energy Source?

66You might say, do we really have enough sunlight? The answer is we have 10,000 times more than we need. . . . If we captured one part in 10,000 of the sunlight, we would meet all of our energy needs.99

—Ray Kurzweil, scientist, inventor, and futurist.

66Wind and solar power are enormously appealing as planet-friendly sources of energy—but those who think we can completely rely on them in the future are dreaming.99

—Simon Grose, science editor of the Australian newspaper *Canberra Times*.

When people consider solar power as an energy source, they often think that it is only realistic in areas that have nearly constant sunlight. Germany, however, has disproved this perception. It is a country that is known for cloudy skies and year-round rainy weather. Many German towns have only about 1,500 hours of sunshine per year, which is about half as much as Spain and Portugal. But in spite of its climate hurdles, Germany has become a global leader in solar power, generating an estimated 55 percent of the world's total photovoltaic energy. One example is the Waldpolenz solar park, which is scheduled for completion by the end of 2009 and will be the largest solar power station in Germany. Located at a former military air base in the townships of Brandis and Bennewitz in eastern Germany, the energy park covers 272 acres (110 ha), an area that is equivalent in size to about 200 soccer fields. It features 550,000 photovoltaic modules, and when it is fully operational it will be one of the largest photovoltaic projects that have ever been constructed. Matthias Willenbacher, one of the general contractors of the Waldpolenz project, explains the significance of the solar park and the impact he believes it will have on the demand for solar energy: "No other solar power plant in the world is as big and as cost-effective as the . . . project in Brandis. Within just a few years the price of solar electricity produced on your own rooftop will

> " When people consider solar power as an energy source, they often think that it is only realistic in areas that have nearly constant sunlight. Germany, however, has disproved this perception. "

be cheaper than the power supplied by the energy utilities. Photovoltaics will then reach completely new dimensions because everyone will want their own installation. That will launch an unprecedented boom."[12]

The Rise and Fall of Solar Power in America

Although the United States trails Germany and Japan in utilizing solar power, this was not always the case. During the late 1970s, when fuel prices were skyrocketing and people feared energy shortages, President Jimmy Carter established the Solar Energies Research Institute and allocated $3 billion to help jump-start solar energy technology. America began to emerge as a global solar power leader, with citizens and entrepreneurs embracing the technology and installing PV panels and solar hot water heating systems. Carter himself even had panels installed on the White House roof in order to make a public statement about his administration's commitment to a clean, renewable energy source that would help to free America from its dependence on foreign oil. The solar industry began to grow, with the number of solar-collector manufacturing firms steadily rising. Then Carter lost his reelection bid to Ronald Reagan in

1980, and the president-elect did not consider alternative energy to be a priority. The solar panels on the White House were removed, and Reagan's administration began to disassemble the solar movement piece by piece. Denis Hayes, who was hired by Carter to lead the solar power initiative, explains: "In June or July of 1981, on the bleakest day of my professional life, they descended on the Solar Energy Research Institute, fired about half of our staff and all of our contractors, including 2 people who went on to win Nobel prizes in other fields, and reduced our $130 million budget by $100 million."[13]

By 1986 the Reagan administration had slashed funding for solar energy research and programs and substantially reduced the federal tax credits that had previously been given to those who installed solar water heating. This had an effect not only on solar power growth in the United States but other countries as well. Travis Bradford, president and founder of the Prometheus Institute for Sustainable Development, explains the ramifications of Reagan's decisions: "The message to the renewable- and solar-energy communities was clear, and the industry came to a virtual halt not only in America but throughout the world. America represented almost 80 percent of the world market for solar energy at that time, and when research funds in America dried up, the remaining governments of the industrialized world followed in step."[14] The solar power market continued to lose ground during the late 1980s and early 1990s as oil prices fell, which further diminished people's motivation to cultivate renewable energy sources. According to Bradford, this resulted in "another two decades of inertia for the global solar-energy industry."[15]

The Resurgence of Solar Power

When oil prices began to climb in the 1990s, solar power again emerged as a viable alternative energy source, and Japan and Germany soon became world leaders in solar power development. Because both countries experience low sunlight, they were overly dependent on expensive foreign sources of energy and were highly motivated to develop lower-cost, renewable power sources. Japan implemented the New Sunshine Program in 1993, which expanded the original Sunshine and Moonlight energy conservation and technology projects. In 1999 the German government implemented its 100,000 Roofs Program, which includes 10-year low-interest loans for PV installation, and its Electricity Feed-in

Law also took effect that same year. The latter features generous incentives for German farmers, homeowners, and industrialists to install solar power and connect their systems to the country's electrical grid. In return, consumers are reimbursed by power companies for their electricity usage. Today in Germany more than 300,000 PV systems are in use, which is triple the government's original goal, and the solar power that they generate is 1,000 times more than what was generated in 1990.

Spain, which has also embraced solar technology, is in an ideal location because it has more available sunshine than any other country in Europe. Some of Spain's major solar power facilities are located in the cities of Seville, Salamanca, and Lobosillo, and the world's largest rooftop solar power installation is on the General Motors plant in Zaragoza. It features 85,000 solar panels that cover 2 million square feet (186,000 square m) of roof. Another Spanish solar installation—a rather unusual one—is located in Santa Coloma de Gramenet, which is outside Barcelona. Town officials wanted to utilize solar power, but the town is densely built up, with 124,000 people residing on just 1.5 square miles (4 square km), so enough flat, open land was not available. They solved the problem by placing 462 photovoltaic panels on top of mausoleums in a cemetery. Today, these panels provide electricity for about 60 homes, and officials plan to expand the installation to generate even more power.

> " By 1986 the Reagan administration had slashed funding for solar energy research and programs and substantially reduced the federal tax credits that had previously been given to those who installed solar water heating. "

The United States still lags behind Spain, Germany, and Japan in its solar power utilization, but that is beginning to change. According to the SEIA, both commercial and residential PV markets grew significantly during 2007. PV manufacturing grew by 74 percent in the United States and PV installations increased by 45 percent, both of which were among the fastest growth rates in the world. The SEIA explains: "In 2007, the U.S. solar energy industry saw a glimpse of a gigawatt future. . . . Thousands of U.S. jobs were created and billions of dollars were invested. And, the industry strengthened its presence in Washington and our united coalition support across the country."[16] The Solar Energy Research Institute, which suffered from drastic funding cuts during the Reagan administration, is now the Golden, Colorado–based National Renewable Energy Laboratory (NREL), the primary U.S. agency for renewable energy and energy efficiency research and development. The NREL receives nearly $380 million per year in funding from the federal government and is dedicated to advancing the use of solar power and other renewable energy sources in homes, commercial buildings, and vehicles.

Solar Light Versus Solar Heat

For a number of years PV has been the most prevalent form of technology used to capture solar energy and convert it into electricity, as well as the most widely recognized. Science writer Carolyn Gramling explains: "When it comes to solar power, many people are most familiar with photovoltaic cells. . . . In addition to providing electricity, hot water and heat to homes and businesses, photovoltaic systems supply power to everything from ocean weather buoys and communications equipment to streetlights and satellites."[17] The major downside of PV is its cost, because in terms of price per kilowatt hour, it is much more expensive to generate electricity with PV technology than to produce it at coal-fired power plants. This is largely because of the high cost of raw materials (such as copper and high-grade silicon) that are used to manufacture solar cells. PV costs are steadily declining and are expected to continue to decrease over the coming years, but price is still often a deterrent to building large PV power plants.

> " Spain, which has also embraced solar technology, is in an ideal location because it has more available sunshine than any other country in Europe. "

Recently, interest in large-scale concentrating solar thermal operations has been renewed because they can generate electricity at a cost of 10 to 15 cents per kilowatt hour, compared to PV at about 30 cents per kilowatt hour. This is because solar thermal plants are built with basic materials such as steel, glass, and concrete, all of which are common and readily available. Another reason solar thermal operations are attractive is that in addition to generating electricity, they can also store excess energy for use when it is needed, such as when it is nighttime or when skies are cloudy. Virginia Lacy writes: "With energy storage, some solar thermal power designs are projected to be able to generate a steady stream of power for up to 20 hours in a given day."[18]

> Recently, interest in large-scale concentrating solar thermal operations has been renewed because they can generate electricity at a cost of 10 to 15 cents per kilowatt hour, compared to PV at about 30 cents per kilowatt hour.

Large concentrating solar thermal operations use several different methods of focusing and capturing the sun's heat. One method is parabolic troughs, which are long, curved mirrors that are arranged side by side. These troughs capture and focus the sun's energy toward pipes that contain synthetic oil. As the constantly circulating oil absorbs solar energy it grows hotter, creating high-pressure superheated steam that powers an electricity-producing turbine. California's nine Solar Electric Generating Stations (SEGS), located at three massive facilities in the Mojave Desert, feature parallel rows of over 1 million parabolic troughs that cover more than 1,600 acres (647 ha). Together, the nine SEGS produce enough electricity to power half a million homes, and they represent the largest solar thermal operation in the world.

Stored Solar Power

A vast amount of the sun's energy is stored in organic matter such as wood, agricultural crops and other vegetation, animal manure, and decomposing waste. Known as biomass, this material is a source of stored solar energy that, like other solar power sources, is renewable because its supplies are not limited—trees and crops can continuously be grown, and waste will always exist. According to the NEED Project, wood (including logs, woodchips, bark, and sawdust) comprises about 65 percent of all biomass energy, while agricultural waste products, garbage, and landfill gas make up the rest. When biomass is burned, the stored energy from the sun is released in the form of heat.

One of the main uses for biomass is the generation of electricity. Power plants that burn biomass, called waste-to-energy plants, burn waste products instead of coal to produce superheated steam, which is then captured and used to create electricity. One example is the Baltimore Refuse Energy Systems Company (BRESCO) in Baltimore, Maryland, which has the capacity to burn more than 2,250 tons (2,040 t) of trash per day, and generates more than 500,000 pounds (227,000 kg) of steam per hour. Some of the steam is sent through underground pipes to Trigen Energy Corporation for use in heating or cooling homes and office buildings, and the rest is used to generate electricity, which is sold to customers and is enough electrical energy for an estimated 40,000 homes.

> **Another way of using biomass is to recycle the methane gas that is emitted from decaying garbage at landfills.**

Another way of using biomass is to recycle the methane gas that is emitted from decaying garbage at landfills. According to a 2008 study by the Environmental Protection Agency, Americans generate roughly 30 million tons (27 million metric tons) of food waste each year, and an estimated 98 percent of that waste ends up in landfills. After it is buried, it begins to decompose beneath the ground, and methane is eventually released into the atmosphere. Because methane is a flammable gas that is highly explosive, landfills are required to have methods in place to collect it and dispose of it. Only about 16 percent of the estimated 2,300 landfills in the United States recycle the gas, while the rest just burn it off. As a result, writes Frederick Reimers, "the potential electricity literally goes up in smoke."[19] Waste Management, Inc., a major landfill owner and operator, has agreements with companies throughout the United States to produce methane gas and help power their operations. By 2012, Waste Management hopes to create enough power from landfills nationwide to power about 700,000 homes.

A Solar Power Renaissance?

From its rise and fall during the 1970s and 1980s to its resurgence today, solar power has become a widely sought source of energy. It has the capacity to heat water, heat and cool buildings, and generate electricity, although its true potential has barely been tapped. As scientists continue to study solar power, they are beginning to overcome its hurdles, such as a price that is higher than energy produced from burning coal. As technology becomes even more sophisticated, solar power will undoubtedly become an even more viable energy source. As Bradford writes: "Although it will be many years before solar energy provides a substantial amount of the world's energy generation, awareness of the inevitability of the solar solution will have a surprisingly dramatic impact on electric utilities, government policy makers, and end users much sooner than most predict."[20]

Is Solar Power a Viable Energy Source?

> **Scientists have confirmed that enough solar energy falls on the surface of the earth every 40 minutes to meet 100 percent of the entire world's energy needs for a full year. Tapping just a small portion of this solar energy could provide all of the electricity America uses.**

—Al Gore, "Al Gore's Speech on Renewable Energy," NPR.com, July 17, 2008. www.npr.org.

Al Gore is a former vice president of the United States, an author, and a spokesperson on behalf of environmental issues.

> **The solar fraud is the litany of unrealistic, rosy predictions of a solar future. It involves lying with statistics and attempting to manipulate the public through numerous coercive means.**

—Howard C. Hayden, *The Solar Fraud: Why Solar Energy Won't Run the World*. Pueblo West, CO: Vales Lake, 2004, p. v.

Hayden is professor emeritus of physics at the University of Connecticut.

66 The sun—that power plant in the sky—bathes Earth in ample energy to fulfill all the world's power needs many times over. It doesn't give off carbon dioxide emissions. It won't run out. And it's free. 99

—Susannah Locke, "How Does Solar Power Work?" *Scientific American*, October 20, 2008. www.sciam.com.

Locke is a freelance science journalist from New York.

66 Solar energy is the most abundant form of power in the universe. 99

—Roger Golden, "Solar Energy—the Only True Source of Power," Ecoble, September 4, 2008. http://ecoble.com.

Golden is a writer from Jacksonville, Florida.

66 For too long solar energy has had the whiff of elitism about it, as if it were the fanciful product of people whose musings aren't quite grounded in reality. 99

—*Las Vegas Sun*, "Solar as Moneymaker," editorial, March 7, 2008. www.lasvegassun.com.

The *Las Vegas Sun* is a newspaper based in Henderson, Nevada. It is known for its strong editorial positions on various issues.

66 Perhaps the current bleak global energy outlook will catalyze the action needed to overcome the hurdles that have prevented solar energy from becoming a viable alternative energy source. 99

—George Zobrist, "Solar Energy—an Alternative Energy Source," *Today's Engineer*, May 2008. www.todaysengineer.org.

Zobrist is professor emeritus at the University of Missouri–Rolla's Department of Computer Science.

> **66** Clean energy is no longer the promise of the future but the reality of today: the technologies work, and also save money—sometimes tens of thousands of dollars on a school, business, or municipal electric bill. **99**

—Massachusetts Technology Collaborative, "Clean Energy Opportunities," September 2007. http://masstech.org.

Massachusetts Technology Collaborative is the state's development agency for renewable energy.

> **66** It's a myth these solar panels. It's a great idea in the future, we're not there yet. I had them on my house in Sacramento, electric bills were supposed to go down, I didn't notice diddly-squat. I didn't even know if the things worked. **99**

—Rush Limbaugh, "The Solar Panel Myth," *Rush Limbaugh Show*, January 28, 2007. www.rushlimbaugh.com.

Limbaugh is a conservative talk show host.

> **66** The sun's heat and light provide an abundant source of energy that can be harnessed in many ways. **99**

—National Renewable Energy Laboratory (NREL), "Solar Energy Basics," June 25, 2008. www.nrel.gov.

NREL focuses on renewable energy and energy efficiency research and development.

> **66** While the U.S. is still a major player in solar research, it has fallen behind in reaping profits from solar cells. **99**

—Otis Port, "Another Dawn for Solar Power," *BusinessWeek*, September 6, 2004. www.businessweek.com.

Port is a senior writer for *BusinessWeek* magazine.

Is Solar Power a Viable Energy Source?

- According to scientists Ken Zweibel, James Mason, and Vasilis Fthenakis, the energy in sunlight striking Earth for **40 minutes** is equivalent to worldwide energy consumption in a year.

- Solar power currently provides less than **.1 percent** of the electricity generated in the United States.

- The U.S. Department of Energy states that covering less than **0.2 percent** of the planet with **10-percent**-efficient solar cells would provide twice the power used worldwide.

- According to the National Renewable Energy Laboratory, solar cells built in the 1950s had efficiency ratings of less than **4 percent**, while today a typical photovoltaic cell has **15 percent** efficiency.

- Large-scale concentrating solar thermal operations can generate electricity at a cost of **10 to 15 cents** per kilowatt-hour, compared to PV at about **30 cents** per kilowatt-hour.

- During 2007 PV manufacturing in the United States grew by **74 percent**, and PV installations grew by **45 percent**, both among the fastest growth rates in the world.

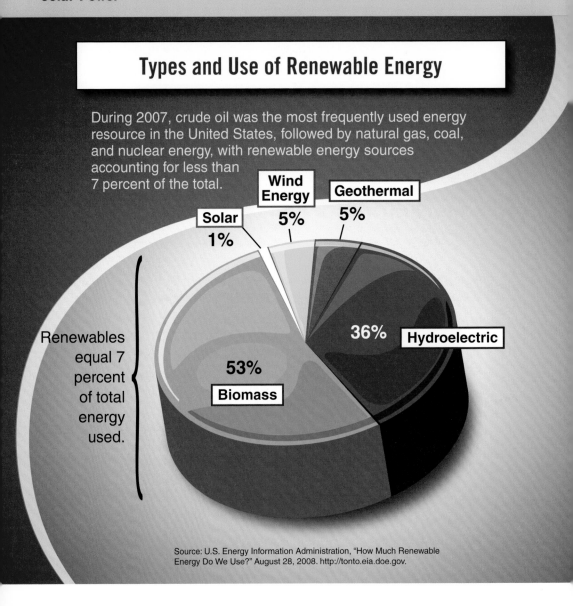

Types and Use of Renewable Energy

During 2007, crude oil was the most frequently used energy resource in the United States, followed by natural gas, coal, and nuclear energy, with renewable energy sources accounting for less than 7 percent of the total.

Solar 1%

Wind Energy 5%

Geothermal 5%

36% Hydroelectric

Renewables equal 7 percent of total energy used.

53% Biomass

Source: U.S. Energy Information Administration, "How Much Renewable Energy Do We Use?" August 28, 2008. http://tonto.eia.doe.gov.

- Although Germany has much less sunlight than the United States, the country installs an estimated **8 times** as much PV because of government incentives that stimulate the demand for solar power.

- During 2006 the state of California implemented the largest solar power operation outside of Germany by enacting the **California Solar Initiative** and the **Million Solar Roofs** bill.

Solar Power Potential: Germany and the United States

The United States has enormous potential for solar power, especially in the southwestern deserts, which have some of the best solar resource levels in the world. Yet even though Germany has fewer solar power resources, because of government incentives the country installs eight times as much photovoltaic technology as the United States. These maps show how the two countries compare in terms of available solar resources and potential power generation.

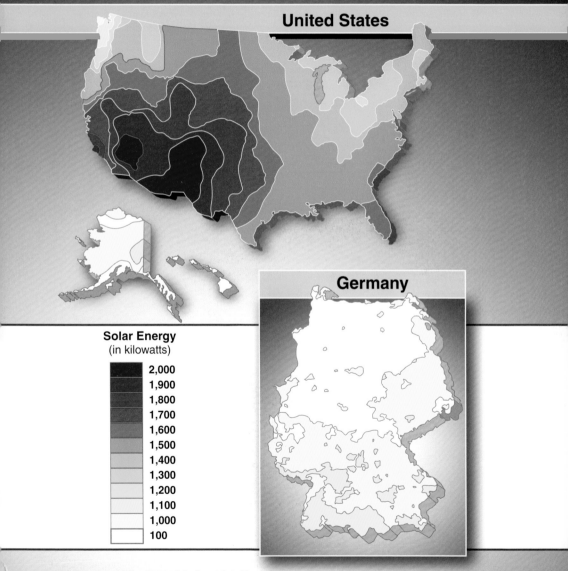

United States

Germany

Solar Energy
(in kilowatts)

	2,000
	1,900
	1,800
	1,700
	1,600
	1,500
	1,400
	1,300
	1,200
	1,100
	1,000
	100

Source: Solar Energy Industries Association (SEIA) and Prometheus Institute, *US Solar Industry: 2007 Year in Review*, 2008. www.seia.org.

World Opinions on Energy

A poll by WorldPublicOpinion.org shows that people throughout the world strongly support alternative energy sources, including solar power. The survey involved nearly 21,000 participants from 21 countries.

"I would like you to consider different ways to deal with the problem of energy. For each one please tell me if you think our country should emphasize it more, less, or the same as now."

Emphasize more   Same as now   Emphasize less   Don't know/not sure

Installing solar or wind energy systems
77%   8%   8%
7%

Modifying buildings to make them more energy efficient
74%   8%   11%
7%

Building nuclear energy power plants
40%   17%   30%   13%

Building coal- or oil-fired power plants
40%   17%   33%
10%

0%   20%   40%   60%   80%   100%

Percentage

Note: Responses represent the average of participants from all 21 countries.

Source: WorldPublicOpinion.org, "World Public Opinion on Energy," November 19, 2008. www.worldpublicopinion.org.

The Growth of Renewable Energy

Since 1980 many countries have become more reliant on solar power and other renewable energy sources for their energy needs. This graph shows those with the highest increases in renewable energy consumption from 1980 to 2005.

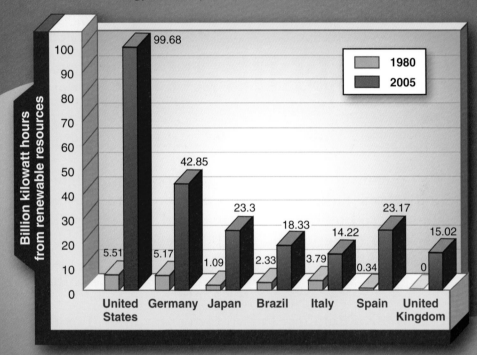

Source: U.S. Energy Information Administration, "International Energy Annual 2005," September 13, 2007. www.eia.doe.gov.

- The world's largest solar thermal operation is the Solar Electric Generating System (SEGS) in California's **Mojave Desert**, which comprises 9 solar power plants with 1 million parabolic troughs and covers more than 1,600 acres (647 ha).

- About **30 percent** of the solar power that reaches Earth is consumed by the hydrologic cycle, which is the continuous circulation of water.

Could Solar Power Replace Fossil Fuels?

66Well-meaning scientists, engineers, economists and politicians have proposed various steps that could slightly reduce fossil-fuel use and emissions. These steps are not enough. The U.S. needs a bold plan to free itself from fossil fuels.**99**

—Ken Zweibel, James Mason, and Vasilis Fthenakis, scientists who are noted authorities on solar power.

66I'm all for efficiency, I'm all for renewables, I'm all for alternatives. But we have to be realistic. That's the point where I part company with all these people who are cheerleaders for 'Oh, we'll quit using oil.' Well, no we won't.**99**

—Robert Bryce, journalist and the author of *Gusher of Lies: The Dangerous Delusions of "Energy Independence."*

Crude oil, coal, and natural gas would never have existed without the sun. When these fossil fuels formed hundreds of millions of years ago, the living plants and animals from which they were created depended on the sun's energy for life. After the organisms died, decomposed, and were buried deep beneath Earth's surface, the fuels that they evolved into over time represented an ancient, stored source of sunlight.

Today, people all over the world depend on fossil fuels for heating, cooling, electricity generation, hot water, and cooking, as well as fuel for cars, buses, trucks, and all other forms of transportation. In fact, fossil fuels supply more than 80 percent of energy that is currently in use worldwide. Although solar power has already proved to have tremendous potential as an energy source, for it to completely replace fossil fuels would require a long-term, complex, and extremely expensive undertaking, as Scripps Institution of Oceanography Meteorologist Richard Somerville, explains: "There won't be a single silver bullet to wean the world from fossil fuels, because they contribute some 80% of the present-day global energy supply, so the task ahead is enormous."[21] Yet even with the immense challenges that must be overcome, Somerville and many other scientists believe that the world's primary future energy source is solar power and other sources of renewable energy, rather than fossil fuels.

> After the organisms died, decomposed, and were buried deep beneath Earth's surface, the fuels that they evolved into over time represented an ancient, stored source of sunlight.

Evolution of Fossil Fuel Use

It is widely believed that humans first discovered crude oil hundreds, or perhaps even thousands, of years ago. Ancient peoples living in Middle Eastern countries, which have historically been rich in oil resources, noticed the mysterious black substance seeping out of cracks in the ground. When the oil reached the surface, its liquids vaporized into the air, leaving behind lumps of a soft, sticky, oil by-product known as bitumen, or pitch. Because the substance was believed to have medicinal powers, it was often used as a dressing for the wounds of people and animals. The ancient Egyptians used pitch as mortar to build their famed pyramids, as well as for boat building to make the vessels watertight. North American natives used pitch to waterproof their canoes and as fuel for burning ceremonial fires, and during the Revolutionary War the native peoples taught Ameircan troops that pitch could be used to treat frostbite.

Natural gas was a mysterious, and somewhat frightening, phenomenon for those who first encountered it. Sometimes when the colorless, odorless gas escaped from Earth's crust, lightning strikes would cause it to burst into flames. Among the most famous of these natural gas fires occurred on Mount Parnassus in ancient Greece in approximately 1,000 B.C. A goatherd reportedly came across what he called a "burning spring,"[22] which was actually a gas flame rising from a fissure in the rock. Believing this flame to be a message from the gods, the Greeks built a sacred temple on the site of the fire. A priestess known as the oracle of Delphi lived in the temple and delivered prophecies that she claimed had been inspired by the divine flame.

> **Sometimes when the colorless, odorless gas escaped from Earth's crust, lightning strikes would cause it to burst into flames.**

Coal was also discovered and used during ancient times, as the DOE writes: "Coal has been used for heating since the cave man. Archeologists have also found evidence that the Romans in England used it in the second and third centuries (100–200 AD)."[23] During the 1300s the Hopi Indians, who were living in the area that is now the southwestern United States, used coal for heating, cooking, and to bake the pottery they shaped from clay.

When European settlers began to arrive in North America during the 1600s, they rediscovered coal but probably did not realize its full potential because they used it infrequently, preferring instead to use wood for fuel. Then during the late 1700s and early 1800s, when the Industrial Revolution began in Europe, the use of coal began to soar. Textile factories and other manufacturing operations throughout Europe used coal to generate power, and ships and locomotives, which had previously burned wood to fuel their steam-powered boilers, began to use coal instead.

When the Industrial Revolution expanded to the United States during the mid-1800s, coal was burned in the huge blast furnaces of factories that produced steel used for weapons during the Civil War. Coal was also used in many other types of manufacturing, as well as to heat homes and businesses. Once considered to be a secondary fuel, coal—which was plentiful as well as cheap—had become the world's primary energy source.

Throughout the years, crude oil, coal, and natural gas have allowed civilization to progress in a way that would not have been possible if they had never been discovered. New York University physics professor Martin Hoffert explains: "We're really lucky—it's almost as if someone wanted us to develop civilization by giving us these fossil fuels! Without them, we would never have had the Industrial Revolution, and would still be living the way people did in the Middle Ages."[24]

A Finite Resource

As crucial as fossil fuels are to modern civilization, they are a nonrenewable, finite resource—which means that sometime in the future they will all be depleted. Exactly what quantities of these fuels are left, or how long they will last, is often a topic of spirited debate. Scientists are able to make relatively accurate quantity estimates by using seismic technology, in which explosions either on the surface or underground generate sound waves that reflect off the rock layers in the earth that hold fossil fuels. The National Energy Foundation states that the rate at which the sound waves are reflected creates a picture of the underground geology, which helps scientists determine the existence of the fuels. Hoffert adds that the most abundant of all fossil fuels is coal, and the United States still has a large supply of it, as he explains: "Since there is much less oil and natural gas, those fuels will be used up during the 21st century, while scientific estimates show that there is enough coal to last for a few hundred more years."[25]

> **Throughout the years, crude oil, coal, and natural gas have allowed civilization to progress in a way that would not have been possible if they had never been discovered.**

According to a 2008 report by British Petroleum (BP) Worldwide, global energy use has steadily grown over the past decade, and it continues to increase every year. During 2007 coal consumption grew by 4.5 percent over 2006, and coal was the world's fastest-growing fuel for the fifth year in a row. Yet this increased usage is even more striking when comparing consumption over the decade 1997 to 2007. During that time, worldwide coal consumption rose by nearly 37 percent, with the greatest usage spike in developing countries such as China and India. Also between 1997 and 2007 global crude oil and natural gas consumption markedly spiked, with each increasing by more than 30 percent over 10 years. "The problem," says Hoffert, "is that we're using fossil fuels up a million times faster than nature made them. It took hundreds of millions of years for these reserves to form, and as the worldwide demand for energy continues to grow, they're being removed from the ground at a rapid rate—and when they're gone, they're gone."[26]

> " According to a 2008 report by British Petroleum (BP) Worldwide, global energy use has steadily grown over the past decade, and it continues to increase every year. "

Robert Bryce, the author of *Gusher of Lies: The Dangerous Delusions of "Energy Independence,"* is a supporter of renewable energy, including solar power. Bryce disagrees, however, that it will completely replace fossil fuels, and he does not believe that the United States should strive to become self-sufficient in terms of energy resources, as he explains: "The global oil market is just that—it is global. And the idea that we, as the U.S., will be able to divorce ourselves from that global market is ludicrous. In 2005 the U.S. bought crude oil from 41 countries, we bought gasoline from 46 and jet fuel from 26. It's a global market—get used to it."[27] Bryce adds that fossil fuels are what drives the worldwide economy and people will not quit using them for at least 30 to 50 years.

The Transition to Solar Power

Many of the scientists and energy experts who are most enthusiastic about solar power have studied its potential in depth, and some have taken the time to develop detailed strategies for how the world could be weaned from fossil fuels. David R. Mills and Robert G. Morgan, who are with the solar technology company Ausra, performed a modeling study that they published in July 2008. By their estimates, the greatest potential is with solar thermal power, which they say is the most cost-effective way to generate electricity, and the most economical method of storing energy from the sun. Mills and Morgan are convinced that solar power could replace most fossil-fueled electricity, and they contend that large-scale solar thermal systems could greatly reduce America's reliance on oil that is currently imported from foreign countries. This, they say, would result in substantial savings that would more than compensate for the financial investment that is needed to implement these systems. Ken Zweibel, James Mason, and Vasilis Fthenakis have also studied how a transition to solar power would best be accomplished. They have developed what they call a "Solar Grand Plan" in which they describe how the United States could free itself from much of its dependence on fossil fuels by the year 2050. They estimate that this would require a financial investment on the part of the U.S. government of $400 billion, which is substantially less than the Mills and Morgan projection of as much as $6 trillion. But Zweibel, Mason, and Fthenakis also say that the investment in solar technology would be well worth the expense because the payoff is so much greater than the necessary expenditures. They write: "Solar plants consume little or no fuel saving billions of dollars year after year. The infrastructure would displace 300 large coal-fired power plants and 300 more large natural gas plants and all the fuels they consume. The plan would effectively eliminate all imported oil, fundamentally cutting U.S. trade deficits and easing political tension in the Middle East and elsewhere." They add that in addition to reducing dependence on fossil fuels, their plan would open up new opportunities for employment: "Some three million new domestic jobs—notably in manufacturing solar components—would be created, which is several times the number of U.S. jobs that would be lost in the then dwindling fossil-fuel industries."[28]

> " Embracing solar power as the primary energy source will no doubt be a difficult task that will take many years and will require overcoming daunting challenges, including substantial financial investments. "

The Solar Grand Plan calls for energy farms to be constructed in the American Southwest, where photovoltaic panels and concentrating solar heating troughs would be placed on an estimated 46,000 square miles (119,000 square km) of land. Although that is definitely a sizable area, according to Zweibel, Mason, and Fthenakis, the space that would be utilized only represents about 19 percent of the total available southwestern land. Another important element is the construction of an electric current "backbone" that would be capable of sending electricity over high-voltage lines to compressed air storage facilities that would be located throughout the United States. The storage facilities would be underground in caverns, abandoned mines, aquifers, and wells that have been depleted of natural gas. According to data from the natural gas industry and the Electric Power Research Institute, these sorts of geologic formations exist in 75 percent of the United States and are often close to metropolitan areas that typically require large volumes of electricity.

Of all the challenges that exist for transitioning away from fossil fuels, Zweibel, Mason, and Fthenakis say that contrary to what is often assumed about solar power, the biggest challenge is neither technology nor money. They share their thoughts on what the real problem is: "It is the lack of public awareness that solar power is a practical alternative—and one that can fuel transportation as well. Forward-looking thinkers should try to inspire U.S. citizens, and their political and scientific leaders, about solar power's incredible potential. Once Americans realize that potential, we believe the desire for energy self-sufficiency and the need to reduce carbon dioxide emissions will prompt them to adopt a national solar plan."[29]

Embracing solar power as the primary energy source will no doubt be a difficult task that will take many years and will require overcoming daunting challenges, including substantial financial investments. The only way that such a transition can be accomplished is with strong commitment and cooperation among governments, scientists, energy companies, and private citizens, all working together to wean the world from fossil fuels that people are totally dependent on today. Can this be done? Is it realistic to rely on solar energy to replace the fossil fuels that have powered the world for so many years in such diverse ways? Some are highly skeptical about whether this goal is attainable, while solar power proponents argue that it cannot be a matter of "if," but rather when, where, and how. They are convinced that with global energy use continuing to climb every year, and fossil fuels rapidly being depleted, the world has no other choice.

Primary Source Quotes*

Could Solar Power Replace Fossil Fuels?

66 In less time than has passed since the founding of Jamestown, today's coal reserves will be forever gone. Also, most scientists agree that the use of fossil fuels is profoundly altering both local environments and the climate of the world itself. 99

—John C. Mankins, "Energy from Orbit," *Ad Astra*, Spring 2008. www.nss.org.

Mankins is president of ARTEMIS Innovation Management Solutions and a recognized leader in space systems and technology innovation.

66 The sun is the ultimate source of energy. All the energy stored in the earth's reserves of coal, oil, and natural gas is equal to the energy from only 20 days of sunshine. 99

—Sierra Club, "Clean Power Comes on Strong: The Sun," 2008. www.sierraclub.org.

The Sierra Club is one of the most influential environmental organizations in the United States.

* Editor's Note: While the definition of a primary source can be narrowly or broadly defined, for the purposes of Compact Research, a primary source consists of: 1) results of original research presented by an organization or researcher; 2) eyewitness accounts of events, personal experience, or work experience; 3) first-person editorials offering pundits' opinions; 4) government officials presenting political plans and/or policies; 5) representatives of organizations presenting testimony or policy.

66 The biggest problem with the sun is that it's not always around. And during the winter months it behaves like an absentee landlord and shows up only for a few hours. 99

—*PC Magazine*, "Solar Cells That Don't Need the Sun," February 1, 2008. www.pcmag.com.

PC Magazine specializes in issues related to information technology, computers, software, and computer science.

66 More solar energy strikes the surface of the earth in one hour than is provided by all of the fossil energy consumed globally in a year. 99

—Massachusetts Institute of Technology (MIT), "Solar Energy: Capturing the Sun," 2008. http://web.mit.edu.

MIT is a science and technology school in Cambridge, Massachusetts.

66 We need to invest in America's future—a green future. We need to invest in solar, wind, geothermal, wave and other green technologies. Our national security depends on it. 99

—Gavin Newsom, "Demand a New Green Economy," *Huffington Post*, October 8, 2008.

Newsom is the mayor of San Francisco.

66 America needs lots of clean, low-cost, secure electricity. Unfortunately, renewable sources don't fill the bill. 99

—Robert J. Michaels, "Opposing View: Renewables Aren't the Answer," *USA Today*, October 20, 2008. http://blogs.usatoday.com.

Michaels is professor of economics at California State University–Fullerton and an adjunct scholar at the Cato Institute.

❝Solar can be used to decrease our overdependence on foreign sources of oil and natural gas.❞

—Solar Energy Industries Association (SEIA) and Prometheus Institute, *US Solar Industry Year in Review 2007*, April 22, 2008. www.seia.org.

SEIA is a national trade association for the solar energy industry, and Prometheus Institute is devoted to accelerating socially beneficial sustainable technologies.

❝The oil, coal, and nuclear industries (and their supporters in Washington) have slowed the transition to a clean energy economy by blocking tax incentives for renewable energy.❞

—Greenpeace USA, "Coal, Oil, and Nukes vs. a Prosperous, Green America: Greenpeace Debate Backgrounder," October 15, 2008. www.greenpeace.org.

Greenpeace is an environmental advocacy organization.

❝Now, I have nothing against solar power or photovoltaic panels. But if they are such a great investment, why do we need to subsidize them?❞

—Jerry Taylor, "Solar-Powered Welfare," Cato@Liberty, January 5, 2007. www.cato-at-liberty.org.

Taylor is a senior fellow at the Cato Institute.

❝One way to think about solar thermal electric power is that it simply replaces a fossil fuel, such as coal, with the sun to power a conventional electricity plant.❞

—Virginia Lacy, "Generating Electricity from the Sun's Heat," Yahoo! Green, January 22, 2008. http://green.yahoo.com.

Lacy is a consultant with the energy and resources team at the Rocky Mountain Institute.

Could Solar Power Replace Fossil Fuels?

- Coal, crude oil, and natural gas are known as **fossil fuels** because they formed over hundreds of millions of years from the fossilized remains of prehistoric organisms.

- The U.S. Energy Information Administration states that during 2005, **86.2 percent** of worldwide energy production was derived from fossil fuels.

- The burning of fossil fuels is a major source of worldwide pollution, including emissions of sulfur dioxide and nitrogen dioxide that mix with precipitation in the atmosphere and lead to **destructive acid rain**.

- According to a 2008 report by British Petroleum (BP) Worldwide, between 1997 and 2007 worldwide coal consumption rose by nearly **37 percent**, while global crude oil and natural gas each increased by more than **30 percent** over the same period.

- New York University physics professor Martin Hoffert estimates that crude oil and natural gas will be depleted by the **end of the twenty-first century** and that coal supplies will last for **several hundred years**.

Electricity Generation in the United States

Although the use of renewable energy resources has markedly increased in the United States, the bulk of the country's electricity is still generated through fossil fuels such as coal and natural gas.

United States Electrical Power Generation—2006

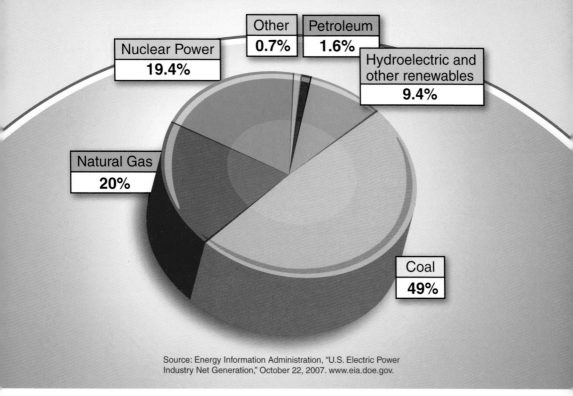

Source: Energy Information Administration, "U.S. Electric Power Industry Net Generation," October 22, 2007. www.eia.doe.gov.

- A 2008 British Petroleum (BP) Worldwide report showed that worldwide coal consumption grew **4.5 percent** between 2006 and 2007 with Chinese consumption accounting for more than **two-thirds** of the growth.

- Scientist/futurist Ray Kurzweil predicts that renewable energy sources will replace fossil fuels within **20 years**.

Worldwide Fossil Fuel Reserves

Crude oil, coal, and natural gas are finite resources, meaning that they will someday be depleted. That is a major reason why scientists and energy experts are especially optimistic about solar power—and its potential is enormous. Some experts say that enough solar energy shines on Earth during a 40-minute period of time to power the entire world economy for a year. A June 2008 report by British Petroleum (BP) shows how fossil fuel reserves vary throughout the world.

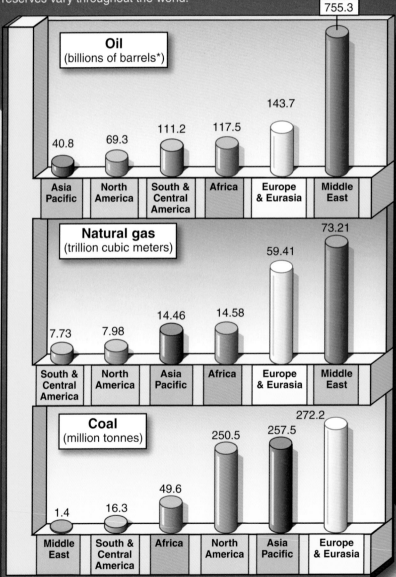

Oil
(billions of barrels*)

Asia Pacific	North America	South & Central America	Africa	Europe & Eurasia	Middle East
40.8	69.3	111.2	117.5	143.7	755.3

Natural gas
(trillion cubic meters)

South & Central America	North America	Asia Pacific	Africa	Europe & Eurasia	Middle East
7.73	7.98	14.46	14.58	59.41	73.21

Coal
(million tonnes)

Middle East	South & Central America	Africa	North America	Asia Pacific	Europe & Eurasia
1.4	16.3	49.6	250.5	257.5	272.2

*Note: 1 barrel = 42 gallons (159 liters)

Sources: British Petroleum, "BP Statistical Review of World Energy," June 2008. www.bp.com; Steve Maxwell, "Solar Will Beat Oil," *Mother Earth News*, September 3, 2008. www.motherearthnews.com.

The World's Top Oil Producers

Scientists are hopeful that solar power will someday replace fossil fuels for virtually every type of energy need, including transportation. Yet crude oil is extremely important because it is currently refined into gasoline and other fuels that are used for all forms of transportation worldwide. This map shows the 15 countries that lead the world in crude oil production.

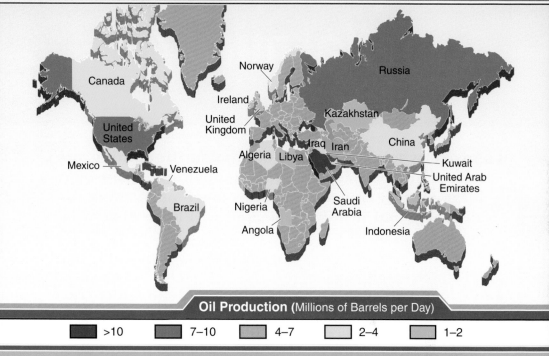

Canada
Norway
Russia
Ireland
United Kingdom
Kazakhstan
United States
Iraq Iran
China
Mexico
Venezuela
Algeria Libya
Kuwait
United Arab Emirates
Brazil
Nigeria
Saudi Arabia
Angola
Indonesia

Oil Production (Millions of Barrels per Day)

| >10 | 7–10 | 4–7 | 2–4 | 1–2 |

Source: Energy Information Administration, "Country Energy Profiles," October 2008. http://tonto.eia.doe.gov.

Could Solar Power Help Stop Global Warming?

> **An overwhelming consensus of scientific opinion now agrees that climate change is happening, is caused in large part by human activities (such as burning fossil fuels), and if left unchecked will have disastrous consequences.**

—Sven Teske, international renewable energy director with Greenpeace.

> **So let's get real and give the politically incorrect answers to global warming's inconvenient questions. Global warming is real, but it does not portend immediate disaster, and there's currently no suite of technologies that can do much about it.**

—Patrick J. Michaels, senior fellow in environmental studies at the Cato Institute.

Whether people agree or disagree about the sun's potential to become the world's leading energy source, one fact cannot be disputed: Unlike energy that is created by burning fossil fuels, solar power is a clean form of energy that does not emit pollutants into the atmosphere. A joint report by the Solar Energy Industries Association (SEIA) and the Prometheus Institute explains: "Solar energy is an emission-free source of electricity and hot water that can be immediately deployed to reduce the nation's growing carbon footprint."[30] The mention of "carbon footprint" refers to CO_2, which is one of the major greenhouse gases that are emitted into the air whenever fossil fuels are burned. These gases are so named because when they are present in the atmosphere, they absorb and hold heat from the sun that has been reflected away from Earth. Since this phenomenon is much the same as how a greenhouse works, it is known as the greenhouse effect, and it is essential for life—without it, the planet would not be able to retain enough heat and would be too cold for living things to survive. Yet even though scientists worldwide understand the importance of the greenhouse effect, many of them are concerned that it is being intensified by human activities, especially the burning of fossil fuels,

> " **Unlike energy that is created by burning fossil fuels, solar power is a clean form of energy that does not emit pollutants into the atmosphere.** "

which pumps billions of tons of heat-trapping gases into the atmosphere each year. They find this alarming for several reasons, one of which is the longevity of the gases. James E. Hansen, director of the Goddard Institute for Space Studies, describes how this relates to CO_2: "Indeed, a quarter of the carbon dioxide (CO_2) that we put in the air by burning fossil fuels will stay there 'forever'—more than 500 years."[31]

Environmental organizations and a growing number of climatologists and other scientists are concerned that greenhouse gases are contributing to a steady increase in the worldwide average temperature that is rising more rapidly than at any other time in history. According to those who are most concerned about global warming, even though the greenhouse gases that are currently in the atmosphere will not disappear for centuries, now is the time to drastically cut back on emissions by aggressively pursuing solar power and other forms of renewable energy. Hansen shares his thoughts:

> Climate change is happening. Animals know it. Many are beginning to migrate to stay within their climate zones. But some are beginning to run out of real estate. They are in danger of being pushed off the planet, to

extinction. Even humans are starting to notice climate change. And they are learning that unabated climate change poses great dangers, including rising sea levels and increased regional climate extremes. Yet the public is not fully aware of some basic scientific facts that define an urgency for action. One stark implication—we must begin fundamental changes in our energy use now, phasing in new technologies over the next few decades, in order to avoid human-made climate disasters.[32]

Crucial Scientific Findings

Although the term "global warming" is well known and highly publicized, knowledge of another term, the "Keeling Curve," is largely confined to the scientific community. Its namesake is the late scientist Charles David Keeling, who during the 1950s became interested in the possibility that the average global temperature was steadily rising due to high levels of carbon dioxide in the atmosphere. He shared the perspective of Swedish scientist Svante Arrhenius, who had written papers during the late 1800s in which he explained that atmospheric concentrations of CO_2 had contributed to long-term variations in climate, and he believed that a likely cause of this was the burning of fossil fuels. Arrhenius's theory was in stark contrast to prevailing scientific beliefs, which held that Earth's vast oceans and forests were capable of absorbing excess CO_2 and thereby had the natural ability to keep the planet's climate in balance. Keeling suspected that this theory was flawed, and he was determined to prove that he was correct.

> **According to those who are most concerned about global warming, even though the greenhouses gases that are currently in the atmosphere will not disappear for centuries, now is the time to drastically cut back on emissions by aggressively pursuing solar power and other forms of renewable energy.**

Armed with little more than a fierce determination, he set up a sampling station at the Mauna Loa Observatory, located on the top of the mountain Mauna Loa on the island of Hawaii. Scientist Andrew Manning, who was later a colleague of Keeling's, explains the importance of that location: "The goal behind starting the measurements was to see if it was possible to track what at that time was only a suspicion: that atmospheric CO_2 levels might be increasing owing to the burning of fossil fuels. To do this, a location was needed very far removed from the contamination and pollution of local emissions from cities; therefore Mauna Loa, high on a volcano in the middle of the Pacific Ocean was chosen."[33]

Using a scientific calibrating instrument known as a manometer that he built himself, and other instruments that he purchased, Keeling began to take measurements in 1958, and he continued his research over the following years. By the 1970s he had made several important discoveries. He found a direct correlation between CO_2 in the atmosphere and fossil fuel burning, and he concluded that only about half of CO_2 concentrations were being absorbed by the oceans—which, as he had long suspected, meant that a substantial amount of CO_2 continued to linger in the atmosphere long after it was emitted. He also determined that CO_2 levels were rising steadily each year, and he plotted this progression on the Keeling Curve. In a paper published in 1993, Keeling described how he had analyzed the increasing amount of CO_2 being emitted into the atmosphere and how he had observed its correlation with the rise in atmospheric CO_2 levels. "The patterns of the rise in CO_2 and of fossil fuel CO_2 emissions are seen to be remarkably similar," he wrote. "This comparison is one of the most convincing indicators that the rise in atmospheric CO_2 is closely related to the injection of CO_2 from fossil fuels."[34] When Keeling first began taking his measurements in 1958, CO_2 levels were estimated to be 315 parts per million (ppm),

> " **[Keeling] found a direct correlation between CO_2 in the atmosphere and fossil fuel burning, and he concluded that only about half of CO_2 concentrations were being absorbed by the oceans.** "

which indicated that for every million molecules of air, 315 molecules of CO_2 were present.

By the time of Keeling's death in 2005, measurements taken at Mauna Loa showed that CO_2 levels had risen to 378 ppm, which represented an increase of 20 percent over 47 years. Other studies showed that CO_2 levels were approximately 385 ppm in 2008. Further studies that involved analyzing bubbles in layers of Antarctic ice sheets indicated that prior to the 1800s, atmospheric concentrations of CO_2 were about 280 ppm—which means that since the Industrial Revolution began, CO_2 levels in the atmosphere have spiked nearly 40 percent. According to the Intergovernmental Panel on Climate Change (IPCC), by the end of the twenty-first century the worldwide concentration of CO_2 is likely to be between 550 and 900 ppm, which the IPCC says would drastically affect world climate.

> **Some scientists believe that global warming does not exist at all and is nothing more than another of Earth's natural climate cycles that have existed throughout history.**

How Serious a Problem Is Global Warming?

Global warming is one of the most controversial and hotly debated issues in science today. Scientists' viewpoints on the topic are highly diverse, ranging from those who warn that worldwide environmental catastrophe is imminent if the warming is not stopped, to others who reject that notion, insisting that the severity of the problem is vastly overblown. Some scientists believe that global warming does not exist at all and is nothing more than another of Earth's natural climate cycles that have existed throughout history. There is even dissension among those who acknowledge that global warming is a reality—some doubt that human actions have actually contributed to it, and others question whether anything should be done to stop it. In a May 31, 2007, interview with National Public Radio, NASA administrator Michael Griffin explained his position on the issue:

I have no doubt that . . . a trend of global warming exists. I am not sure that it is fair to say that it is a problem we must wrestle with. To assume that it is a problem is to assume that the state of Earth's climate today is the optimal climate, the best climate that we could have or ever have had and that we need to take steps to make sure that it doesn't change. First of all, I don't think it's within the power of human beings to assure that the climate does not change, as millions of years of history have shown. And second of all, I guess I would ask which human beings—where and when—are to be accorded the privilege of deciding that this particular climate that we have right here today, right now is the best climate for all other human beings. I think that's a rather arrogant position for people to take.[35]

Many climate scientists disagree with Griffin, insisting that global warming *is* a serious problem, and *must* be addressed immediately. Richard Somerville argues that it is an issue that needs to be a high priority both nationally and internationally, and he adds that people must realize that the costs of neglecting climate change are much larger than the costs of taking action. He writes:

Global climate change is real and serious and ought not to be a partisan issue. The main obstacle to progress is refusing to face reality and dismissing the problem. . . . Climate change is not a problem that the United States or the world can afford to procrastinate about any longer. If we fail to act decisively, the result will inevitably be a severely degraded climate later this century. Different parts of the United States in coming decades will be at risk for rising sea levels, decreased water supplies, altered precipitation patterns, floods, droughts, heat waves, and wildfires. The consequences will affect the domains of public health, economic prosperity, and national security.[36]

Lowering Greenhouse Gas Emissions

Scientists Ken Zweibel, James Mason, and Vasilis Fthenakis opine that as long as fossil fuels are burned in power plants, industrial operations, and in vehicles throughout the world, "millions of tons of greenhouse gases [will continue to be emitted] into the atmosphere annually, threatening the planet." They add that if their Solar Grand Plan were implemented, it would reduce greenhouse gas emissions from power plants by 1.7 billion tons (1.5 billion t) per year, as well as reduce another 1.9 billion tons (1.7 billion t) from hybrid vehicles that would run on solar-generated electricity rather than gasoline. "In 2050," they write, "U.S. carbon dioxide emissions would be 62 percent below 2005 levels, putting a "major brake on global warming."[37]

> " The Waldpolenz energy park in Germany . . . estimates that its solar operation avoids pumping about 25,000 tons (23,000 t) of CO_2 into the atmosphere every year. "

Although the worldwide burning of fossil fuels continues to emit CO_2 and other greenhouse gases into the air, the solar power installations that are currently in operation have proved to be successful in reducing the world's greenhouse gas emissions. The Waldpolenz energy park in Germany, for example, estimates that its solar operation avoids pumping about 25,000 tons (23,000 t) of CO_2 into the atmosphere every year. Waldpolenz's general contractor Matthias Willenbacher explains: "At a time when the whole world is discussing climate change, we are demonstrating the capabilities of renewable energies."[38] In southern Spain, on the outskirts of Seville, a massive solar operation is being expanded that will contain 10 concentrating solar power plants. When the expansion is complete and the Sanlucar la Mayor Solar Platform is fully operational in 2013, it is expected to produce enough electricity to power 180,000 homes and will prevent the emission of about 661,000 tons (600,000 t) of CO_2 into the atmosphere each year.

The Controversy Lingers

Although global warming continues to garner immense publicity and is a controversial topic among scientists, legislators, and environmental organizations throughout the world, the issue remains shrouded in controversy. Scientists on all sides have strong opinions, with some warning of impending environmental catastrophe, and others insisting that the seriousness of the problem has been blown out of proportion. But even if the skeptics are correct—even if global warming is not the dire issue that many say it is—widespread solar power installations, by virtue of their reliance on the sun rather than the burning of fossil fuels, have the potential to drastically reduce atmospheric greenhouse gas emissions. Will this stop global warming? No one can say for sure. But numerous scientists are convinced that the environmental benefits of solar power are more than enough to justify the transition away from fossil fuels.

Could Solar Power Help Stop Global Warming?

66 **Carbon dioxide is foremost among the several greenhouse gases released primarily by fossil-fuel consumption that are radically altering the environment worldwide, including increasing average surface temperatures, changing climate patterns, and sea-level rise.** 99

—Travis Bradford, *Solar Revolution*. Cambridge, MA: MIT Press, 2006.

Bradford is president and founder of the Prometheus Institute for Sustainable Development.

....

66 **Believe it or not, Global Warming is not due to human contribution of Carbon Dioxide (CO_2). This in fact is the greatest deception in the history of science.** 99

—Timothy Ball, "Global Warming: The Cold, Hard Facts?" *Canada Free Press*, February 5, 2007.
www.canadafreepress.com.

Ball is chairman of the Natural Resources Stewardship project in Victoria, Canada, and a former climatology professor at the University of Winnipeg.

....

Bracketed quotes indicate conflicting positions.

* Editor's Note: While the definition of a primary source can be narrowly or broadly defined, for the purposes of Compact Research, a primary source consists of: 1) results of original research presented by an organization or researcher; 2) eyewitness accounts of events, personal experience, or work experience; 3) first-person editorials offering pundits' opinions; 4) government officials presenting political plans and/or policies; 5) representatives of organizations presenting testimony or policy.

66 **If people are presented with a choice to get energy from a polluting source like coal or from a clean source, like solar, it really is a no-brainer. Economically and environmentally, it just makes sense.** 99

—Neal Lurie, interview by Eco-Blogger, "An Interview with Neal from American Solar Energy Society," Boulder Green, April 21, 2008. http://boulder.typepad.com.

Lurie is director of marketing and communications at American Solar Energy Society.

66 **The transition to a clean-energy economy is not some luxury that we can only afford in good financial times.** 99

—Clint Wilder, "Clean Energy Outlook: Hands Off the Panic Button," *Huffington Post*, October 26, 2008. www.huffingtonpost.com.

Wilder is a business and technology journalist who covers the clean energy industry as contributing editor at Clean Edge, a research and online publishing firm.

66 **If the world is warming, it is much more reasonable to adjust to it, rather than try to stop it. If sea levels rise, we can build dykes and move back from the coasts. It worked for Holland.** 99

—John Stossel, "The Global Warming Myth?" April 20, 2007. http://abcnews.go.com.

Stossel is an investigative journalist and author.

66 **Solar technologies diversify the energy supply, reduce the country's dependence on imported fuels, improve air quality, and offset greenhouse gas emissions.** 99

—U.S. Department of Energy Office of Energy Efficiency and Renewable Energy (EERE), "Solar Energy Technologies Program," March 19, 2008. www1.eere.energy.gov.

The EERE seeks to strengthen America's energy security, environmental quality, and economic vitality.

> ❝The sunshine that falls on the Earth in a single hour contains enough energy for all the planet's power needs for a year. By harnessing a tiny fraction of that energy, we can gradually replace the use of coal and gas in electricity power plants, the biggest source of Earth-warming carbon dioxide.❞

—Evan I. Schwartz, "Use the Sun to Cool Down Our Planet," *USA Today*, April 25, 2007. http://blogs.usatoday.com.

Schwartz is producer and writer of *Saved by the Sun*, a television documentary about solar power.

> ❝My opinion is that we need to completely wean ourselves from fossil fuels. By the year 2050 we will have to produce between 100 and 300 percent of all the energy we use right now from sources that do not produce any carbon dioxide—and that's just to *stabilize* the atmosphere.❞

—Martin Hoffert, interview with author, October 11, 2008.

Hoffert is professor emeritus of physics and former chair of the Department of Applied Science at New York University.

> ❝I would simply point out that climate alarmism has become a cottage industry in this country and many others, but a growing number of scientists and the general public are coming around to the idea that climate change is natural and that there is no reason for alarm.❞

—James Inhofe, "Global Warming Alarmism Reaches a 'Tipping Point,'" speech to United States Senate, October 26, 2007. http://epw.senate.gov.

Inhofe is a U.S. senator from Oklahoma and a member of the Environment and Public Works Committee.

66 Energy efficiency and renewable energy technologies are available for large-scale deployment today to immediately begin to tackle the climate change crisis. 99

—Charles F. Kutscher, ed., *Tackling Climate Change in the U.S.*, American Solar Energy Society, January 2007.
http://ases.org.

Kutscher is an engineer and manager of the Thermal Systems Group at the National Renewable Energy Laboratory in Golden, Colorado.

66 What especially worries me is that if anyone dares to question the dogma of the global warming doomsters who repeatedly tell us that C not only stands for carbon but for climate catastrophe, we are immediately vilified as heretics or worse as deniers. 99

—David Bellamy, "The Global Warming Myth," New Zealand Climate Science Coalition, June 2007.
http://nzclimatescience.net.

Bellamy is a professor at the University of Durham in the United Kingdom and an outspoken global warming skeptic.

Could Solar Power Help Stop Global Warming?

- Unlike fossil fuel burning, solar power emits no **carbon dioxide** or other heat-trapping gases into the atmosphere.

- Research has shown that atmospheric levels of CO_2 have risen from **220 parts per million** (ppm) in the preindustrial era to **385 ppm** in 2008; during that same period, Earth's **average temperature has warmed** at what many scientists say is an unprecedented rate.

- According to Solar Energy International, an average of **16 million tons** (14.5 million t) of CO_2 are emitted into the atmosphere every 24 hours by human use worldwide.

- A study announced in November 2008 by the National Coal Council estimates that efficiency upgrades and retrofits to existing coal-fired power plants could avoid about **220 million** tons (200 million t) of CO_2 atmospheric emissions.

- The energy technology company Ausra predicts that converting the U.S. electricity grid to solar thermal power would reduce overall atmospheric greenhouse gases by **40 percent**.

- The American Solar Energy Society states that the **transportation sector** is the source of about **one-third** of carbon emissions in the United States.

The Greenhouse Effect and Global Warming

Carbon dioxide (CO_2), water vapor, nitrous oxide, and methane are called greenhouse gases because they trap and hold solar heat in the atmosphere that is reflected away from the surface of Earth. This "greenhouse effect" is essential for life—without it, the planet would be too cold for living things to survive. But a growing number of scientists are concerned that the greenhouse effect is being intensified by human actions, especially the burning of fossil fuels, which pumps billions of tons of CO_2 and other heat-trapping gases into the atmosphere each year. Solar power is a clean form of energy that emits no greenhouse gases or other pollutants into the air, which means that its use could potentially help slow down or even stop global warming.

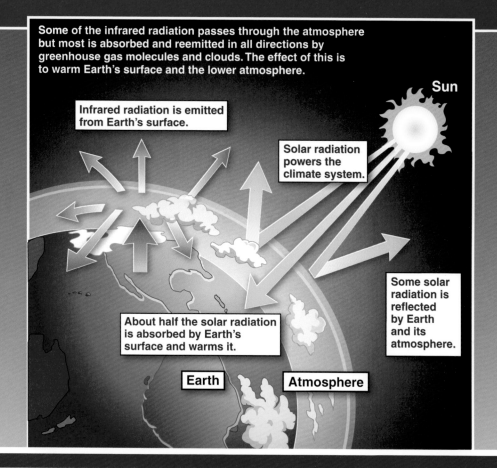

Some of the infrared radiation passes through the atmosphere but most is absorbed and reemitted in all directions by greenhouse gas molecules and clouds. The effect of this is to warm Earth's surface and the lower atmosphere.

Sun

Infrared radiation is emitted from Earth's surface.

Solar radiation powers the climate system.

Some solar radiation is reflected by Earth and its atmosphere.

About half the solar radiation is absorbed by Earth's surface and warms it.

Earth

Atmosphere

Source: Intergovernmental Panel on Climate Change, "What Is the Greenhouse Effect?" March 24, 2008. http://ipcc-wg1.ucar.edu.

Concern About Global Warming Increasing

Global warming is a controversial topic: Many scientists are alarmed about it and urge immediate government action to stop it from getting worse, such as phasing out the use of fossil fuels and relying more on solar power and other renewable energy sources. A joint opinion poll by ABC News, Planet Green, and Stanford University showed that there is much greater awareness of global warming, and concern about its effects, today than was the case during a 1997 survey by Ohio State University. These graphs show how Americans' opinions have changed over the past decade.

How much do you feel you know about global warming?

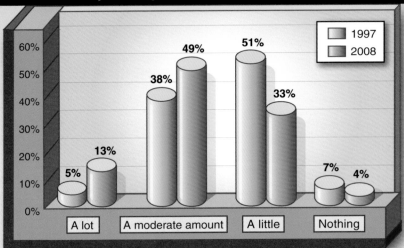

How important is the issue of global warming to you personally?

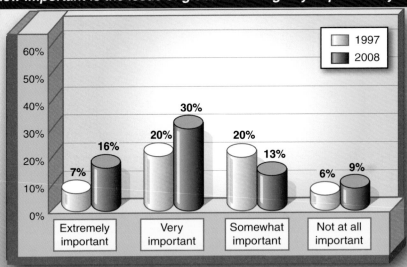

Source: ABC News/Planet Green/Stanford University, "Fuel Costs Boost Conservation Efforts: 7 in 10 Reducing Carbon 'Footprint,'" August 9, 2008. http://abcnews.go.com.

The Steady Rise of CO₂ in the Atmosphere

In 1958 American scientist Charles David Keeling began taking measurements of CO_2 levels in an observatory high atop Mauna Loa on the island of Hawaii, and he continued to do so over the following years. By the 1970s he had established that there was a direct correlation between CO_2 in the atmosphere and fossil fuel burning, and he also determined that CO_2 levels were rising steadily each year. Keeling plotted his findings on what is now known as the Keeling Curve, and since his death in 2005, scientists at the Mauna Loa observatory have continued to record CO_2 levels. This graph shows the progression from March 1958 through November 2008.

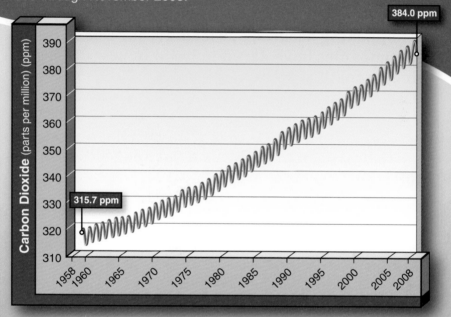

Source: Scripps Institution of Oceanography, "Atmosphere CO₂: Mauna Loa Record," December 2008. http://scrippsco2.ucsd.edu.

- An October 2007 report by the Environmental Protection Agency (EPA) showed that if plug-in hybrid vehicles had a **30 percent** market share in the United States by 2025, and continued to maintain it for 25 more years, this would reduce CO_2 emissions of up to **12 million tons** by 2050.

- The Department of Energy states that during 2006, vehicle emissions of CO_2 totaled **2.2 billion tons** (2 billion t) and electricity generation emissions **totaled 2.4 billion tons** (2.2 billion t).

What Is the Future of Solar Power?

66Solar energy is bound to be a big part of our future. It's happening because our leaders, our scientists and our companies have dared to build in new directions, using strengths we already had.99

—Clark Hughes, editorial page editor of the *Bay City Times* newspaper.

66There is an old adage that if something sounds too good to be true, it probably is. That adage is especially applicable to solar energy.99

—Howard C. Hayden, professor emeritus of physics at the University of Connecticut.

On July 17, 2008, former U.S. vice president Al Gore gave a speech in which he proposed an aggressive goal for America: that the country should commit to producing 100 percent of its electricity from renewable energy within 10 years, which he said was "achievable, affordable and transformative." Gore acknowledged that this would be a challenge for all Americans "in every walk of life: to our political leaders, entrepreneurs, innovators, engineers, and to every citizen."[39] He was aware that many people would reject such a bold notion, but he claimed that the detractors were those who had a vested interest in maintaining the current energy system, even though it was common knowledge that fossil fuels were rapidly being depleted and the damage to the planet from burning these fuels had continued to worsen over the years. He stated: "To those who say the costs are still too high: I ask them to consider whether the costs of oil and coal will ever stop increasing if we keep relying on quickly depleting energy sources to feed a rapidly growing demand all around the world. . . . I for one do not believe our country can withstand 10 more years of the status quo."[40]

Whether Gore's vision of weaning the United States from fossil fuels can be accomplished in such a short timeframe is a contentious issue, because studies have shown that such a transition will be a long-term, complex, expensive undertaking. But aside from when it could conceivably happen, a growing number of scientists and energy experts are convinced that renewable energy, especially solar power, holds tremendous potential for the world's future energy needs and could supply much—if not all—of the worldwide energy that is now largely dependent on fossil fuels.

State-of-the-Art Solar Plants

Numerous solar energy operations throughout the world have already proved to be immensely successful. Some of the world's largest, most powerful solar energy installations are currently in operation, and innumerable others are in the planning phases or under construction. One example is a photovoltaic power plant located in Deming, New Mexico, which is scheduled for completion in 2011 and will be one of the biggest solar facilities in the world. The facility will cover more than 3,200 acres (1,300 ha) of land, generate enough electricity for an estimated 240,000 homes, and provide jobs for as many as 400 people. Also planned for the property will be a $650 million manufacturing facility that will produce and market solar panels.

> **Whether Gore's vision of weaning the United States from fossil fuels can be accomplished in such a short time-frame is a contentious issue, because studies have shown that such a transition will be a long-term, complex, expensive undertaking.**

A Hong Kong–based energy company is constructing a massive solar plant in southeastern Australia, which will be the largest solar project on the continent. The operation will utilize mirror arrays that concentrate light onto advanced high-efficiency photovoltaic cells, and this technology will shrink the required size of the cells, thus lowering the cost. The plant is scheduled to begin operating in 2010 and is expected to continue growing in size until its completion in 2013. When it is fully operational, it will generate enough electricity to power 45,000 homes. Energy officials say that because it represents clean technology, it will keep an estimated 437,000 tons (396,000 t) of greenhouse gas emissions out of the atmosphere each year.

> " A Hong Kong–based energy company is constructing a massive solar plant in southeastern Australia, which will be the largest solar project on the continent. "

Because of the vast solar potential in California's Mojave Desert that is already being realized by the Solar Electric Generating Stations (SEGS), more solar operations are in the planning stages. One, known as the Mojave Solar Park, is being constructed by an Israeli company called Solel and is scheduled for completion in 2011. The installation will cover a massive area—approximately 6,000 acres (2,400 ha), or 9 square miles (23 square km) of desert land. It will use the same parabolic trough technology that currently powers the SEGS. The new solar thermal operation will consist of more than 1 million solar mirrors and will utilize 317 miles (510 km) of vacuum tubing, and its electricity will be ample enough to power 400,000 homes throughout northern and central California. Pacific Gas & Electric's Fong Wan explains the importance of the facility: "The solar thermal project . . . is another major milestone in realizing our goal to supply 20 percent of our customers' energy needs with clean renewable energy."[41]

Futuristic Solar Power

Before his death in 1992, the famed author and scientist Isaac Asimov often spoke about how he had been influenced by a science fiction novel called *The Man Who Awoke*, written in 1933 by Lawrence Manning. The main character in the book was Norman Winters, a rich hermit who invented a potion that put him in a state of suspended animation—and when he woke up, it was in the year 5000. He was shocked to discover that the world had radically changed but not in the way he had expected. Instead of a society brimming with futuristic technology, it was as though the world had gone backward in time, with people living simple lives, walking everywhere they went, and often worrying about where their next meal would come from. No trace of fossil fuels was left because, as the people explained to Winters, prior generations had greedily used up all the resources.

Although the book was a work of fiction, it had an effect on Asimov. He began to devote considerable thought to how fossil fuels were a finite resource that would, inevitably, be depleted, and this heightened his realization of the importance of solar power. In keeping with a concept that had first been suggested in 1968 by scientist Peter Glaser, Asimov proposed the idea of space-based solar power during the 1970s. He envisioned the development of satellites outfitted with arrays of solar cells that could collect the sun's energy in space, convert it into electricity, and then use microwave technology to beam the electrical power back to Earth. By doing this, Asimov opined, solar power could be harvested continuously, without interruption from the planet's atmosphere, clouds, inclement weather, or darkness. He was convinced that space-based solar power was the

> **In 2007 the Pentagon issued a recommendation that the U.S. government take the lead in aggressively pursuing space-based solar technology.**

obvious answer to the looming problem of fossil fuel depletion, as well as the scourge of atmospheric pollution, which he knew was growing worse every year.

Because Asimov had made a name for himself as a writer of science fiction, his concept of space-based solar power must certainly have seemed far-fetched to people when he first proposed it. Today, however, space-based solar power is receiving serious attention. In fact, in 2007 the Pentagon issued a recommendation that the U.S. government take the lead in aggressively pursuing space-based solar technology. Marine Corps lieutenant colonel Paul Damphousse explains the Pentagon's reasoning for support of such an undertaking: "One of the major findings was that space-based solar power does present strategic opportunity for us in the twenty-first century. It can advance our U.S. and partner security capability and freedom of action and merits significant additional study and demonstrations on the part of the United States so we can help either the United States develop this, or allow the commercial sector to step up." Damphousse shared his thoughts on how soon this solar technology might be developed. "While challenges do remain . . . space-based solar power is closer than ever. We are the day after next from being able to actually do this."[42]

> After performing a series of studies, TREC issued a report stating that by using less than 0.3 percent of desert land, thermal solar power would meet a major portion of Europe's electricity needs—at a price that was cheaper than the cost of oil.

Scientists who advocate the space-based solar approach envision that a solar module would orbit Earth around the clock, as the Hubble Space Telescope does today. Because a solar module would be high above the atmosphere, it would receive the full power of the sun's rays, nearly 24 hours a day, 7 days a week, 365 days a year. An article that was published in the Spring 2008 issue of the scientific magazine *Ad Astra* refers to space-based solar power as "inexhaustible energy from orbit" and describes how such a solar power station would work: "Located 22,240 miles [35,800 km] above Earth, a solar power satellite uses mirrored panels to collect continuous energy from the sun. It then redirects the energy onto concentrating photovoltaic arrays, which convert it into electrical power. In turn, this energy is transmitted to the ground and can be routed into an electrical grid as base-load power and ultimately used to light up entire cities."[43]

Tapping the Blazing Sahara Sun

In 2002 German physicist Gerhard Knies began to consider the possibility of building solar power operations in the world's deserts, which are baked in sunlight year-round. He envisioned that these facilities would be the optimal way of generating clean, renewable power that could satisfy the energy needs of many countries throughout the world. In 2003 Knies partnered with the Club of Rome, a global think tank composed of scientists, business people, and current/former heads of state from around the world. The group formed an organization called the Trans-Mediterranean Renewable Energy Cooperation (TREC) and began to develop a project that they called Desertec. Their vision was to build a massive array of concentrating solar power plants in the Sahara Desert that would generate ample amounts of solar power for North Africa and the Middle East, as well as transport solar energy across the Mediterranean Sea to provide electricity for people in Europe.

> **Although challenges still exist, some of them major, no scientific breakthrough in history has been without challenges.**

After performing a series of studies, TREC issued a report stating that by using less than 0.3 percent of desert land, thermal solar power would meet a major portion of Europe's electricity needs—at a price that was cheaper than the cost of oil. TREC projected that the Desertec operation would cost an estimated $400 billion for the many power stations that would need to be constructed in the Sahara. As of 2008, due to a lack of adequate funding and concerns over political instability in the Middle East and North Africa, the Desertec project remained on hold. Knies still remains hopeful that the project will soon be implemented, and he shares his thoughts in an April 2008 press release: "These technologies are well-established and ready to use now. The potential is absolutely colossal. In principle, all of the world's energy needs could be met from less than 1% of the world's desert areas."[44]

Gasoline from Sunshine?

One of the most pressing energy needs throughout the world is crude oil, which is refined into fuel for cars, trucks, and other forms of transportation. Scientists are now beginning to explore ways of generating fuel from sunlight. This is the case with a team of researchers at Sandia National Laboratories in New Mexico, who are working on what they call the Sunlight to Petrol (S2P) project. They have built a prototype device that uses concentrated solar power to chemically "reenergize" CO_2 into carbon monoxide, which could then be used to make hydrogen or serve as a building block for producing "liquid solar fuels" such as methanol, gasoline, and diesel fuel, and possibly even jet fuel. One of the researchers on the S2P project is Ellen B. Stechel, who is manager of Sandia's Fuels and Energy Transitions Department. She notes that scientists have long known that it was theoretically possible to recycle CO_2, but it was often dismissed as impractical or not economically feasible. She explains the renewed interest: "This invention, though probably a good 15 to 20 years away from being on the market, holds a real promise of being able to reduce carbon dioxide emissions while preserving options to keep using fuels we know and love. Recycling carbon dioxide into fuels provides an attractive alternative to burying it."[45]

The Future Beckons

Solar power is an exciting energy concept that has immense potential for the future. Scientists say that the sun's energy has barely begun to be tapped, and as technology becomes more sophisticated, the usefulness of solar power appears to have no limits. Although challenges still exist, some of them major, no scientific breakthrough in history has been without challenges. And just as they have been overcome through technology and scientific perseverance, the same will undoubtedly be true of solar power. Travis Bradford of the Prometheus Institute for Sustainable Development writes: "The inevitability of solar power itself is a powerful concept, and a clear vision of the inevitable will help guide decision making today and in the years ahead. . . . It may be sufficient for now to realize that alternative paths do exist, that the goals of promoting business and the environment need not be mutually exclusive, and that progress toward a practical, sustainable relationship with our planet is not only achievable but inevitable."[46]

What Is the Future of Solar Power?

> 66 I believe that if we really wanted to, we could run the world on solar and wind energy. 99

—Martin Hoffert, interview with author, October 11, 2008.

Hoffert is professor emeritus of physics and former chair of the Department of Applied Science at New York University.

> 66 Wind and solar power are enormously appealing as planet-friendly sources of energy—but those who think we can completely rely on them in the future are dreaming. 99

—Simon Grose, "False Dawn of Solar Power," *Cosmos*, October 25, 2006. www.cosmosmagazine.com.

Grose is science editor of the Australian newspaper *Canberra Times*.

66 The energy challenges our country faces are severe and have gone unaddressed for far too long. Our addiction to foreign oil doesn't just undermine our national security and wreak havoc on our environment—it cripples our economy and strains the budgets of working families all across America. 99

—TheWhite House, "The Agenda: Energy and the Environment," January 2009. www.whitehouse.gov.

The White House Web site presents the Obama administration's positions on energy and many other issues.

66 There is . . . one potential future energy source that is environmentally friendly, has essentially unlimited potential and can be cost competitive with any renewable source: space solar power. 99

—O. Glenn Smith, "Harvest the Sun—from Space," *New York Times*, July 23, 2008. www.nytimes.com.

Smith is a former manager of science and applications experiments for the International Space Station at NASA's Johnson Space Center.

66 I see the opportunities for renewables. I see that they can provide 100 per cent of our energy, and they can be introduced very fast. 99

—Hermann Scheer, interview by Fred Pierce, "Bring on the Solar Revolution," *New Scientist*, May 21, 2008. www.newscientist.com.

Scheer is general chairman of the World Council for Renewable Energy.

66 My vision is that when I fly up and down the state of California that I see everything blanketed, every available space blanketed with solar, if it is parking lots, if it's on top of buildings, on top of prisons, universities, government buildings, hospitals—solar, solar, solar, that is my goal. 99

—Arnold Schwarzenegger, "Governor Schwarzenegger Participates in Launch of New Solar Energy Facility," speech, October 23, 2008. http://gov.ca.gov.

Schwarzenegger is the governor of California.

" The issue with solar is that you're paying high capital costs. But you're avoiding the uncertainties of fuel costs, and you're getting a stream of energy at a fixed priced. As long as we can get in the same ballpark economically, people will demand solar energy, and the government will respond. **"**

—Ken Zweibel, interview by Lindsay Meisel, "An Interview with Solar Power Expert Ken Zweibel," Breakthrough Blog, April 15, 2008. http://thebreakthrough.org.

Zweibel is an international authority on solar power who often speaks on solar policy issues.

" When the day comes that the electricity from solar or nuclear power plants is worth more than the costs associated with generating it, I will be as happy as the next Greenpeace member or MIT graduate to support either technology. **"**

—Jerry Taylor, "Nuclear Energy: Risky Business," *Reason*, October 22, 2008. www.cato.org.

Taylor is a senior fellow with the Cato Institute.

" A confluence of political will, economic pressure and technological advances suggests that we are on the brink of an era of solar power. **"**

—Bennett Daviss, "Solar Power: The Future's Bright," *New Scientist*, December 8, 2007. www.newscientist.com.

Daviss is the author of two books and has written more than 400 articles on education, science, technology, and business.

" Solar installations across just 19 percent of the most barren desert land in the Southwest could supply nearly all of our nation's electricity needs. **"**

—Robert F. Kennedy Jr., "Commentary: Obama's Energy Plan Would Create Green Gold Rush," CNN, August 25, 2008. www.cnn.com.

Kennedy is a senior attorney for the Natural Resources Defense Council and president of the Waterkeeper Alliance.

What Is the Future of Solar Power?

- A 2008 joint report by Clean Edge and Co-op America shows that solar power could meet **10 percent** of the United States' energy needs by 2025.

- A 2008 study by the energy technology company Ausra shows that over **90 percent** of America's electricity grid and vehicle energy needs could potentially be met by solar thermal power.

- A 2007 report by the Pentagon strongly encourages the U.S. government to pursue **space-based solar power**.

- A report released in October 2008 by the European Renewable Energy Council and Greenpeace states that the world could **eliminate fossil fuel use by 2090** by spending trillions of dollars on developing renewable energy technology.

- The U.S. Department of Energy states that its goal is for photovoltaics (PV) to generate enough electrical capacity to power up to **2 million homes, avoid 10 million metric tons** per year of CO_2 emissions, and **employ 30,000 new workers** in the PV industry by the year 2015.

- In 2008 scientists at the Massachusetts Institute of Technology (MIT) announced that they had developed a new method of powering fuel cells that would make solar energy **more affordable** and **easier to store** for homeowners.

Space-Based Solar Power

The potential of solar power is immense, but factors such as necessity for massive areas of open land, storage issues, the ability to transmit energy from desert installations to faraway areas, and lack of availability at night or when skies are overcast, are difficult challenges to overcome. The answer, say many scientists, lies with orbiting satellites that would collect solar power in space and beam it back to Earth. This illustration shows how such futuristic technology would work.

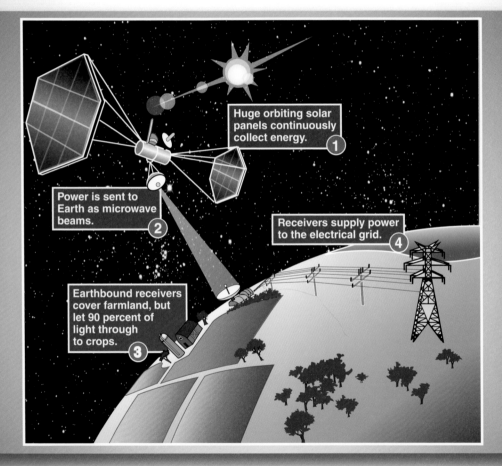

Source: Eric Sofge, "Space-Based Solar Power Beams Become Next Energy Frontier," *Popular Mechanics*, January 2008. www.popularmechanics.com.

Future Potential of Renewable Energies

Even though renewable energy sources are currently used to satisfy only a fraction of the United States' total energy consumption (less than 7 percent in 2007), the U.S. Energy Administration and American Solar Energy Society project that a much larger amount (54 percent) will come from solar and other renewables by the year 2030.

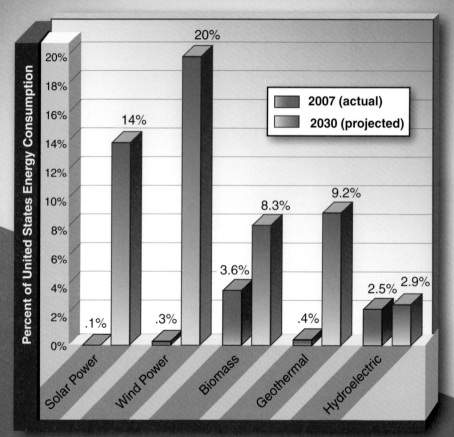

Sources: U.S. Energy Information Administration, "How Much Renewable Energy Do We Use?" August 28, 2008. http://tonto.eia.doe.gov; U.S. Energy Information Administration, "Annual Energy Outlook 2009 Early Release," December 2008. www.eia.doe.gov; Charles F. Kutscher, ed., *Tackling Climate Change in the U.S.*, American Solar Energy Society, January 2007. http://ases.org.

- The U.S. Department of Energy predicts that within **10 years**, photovoltaic power will be competitive in price with traditional sources of electricity.

Americans' Views on Future Energy

During 2008 a national polling firm conducted a survey of 1,000 Americans to ask their opinions on solar and other renewable energies. Participants' responses to two of the questions are shown on these charts.

How important do you think it is for the U.S. to develop and use solar power?

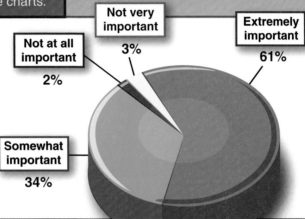

Not very important 3%

Not at all important 2%

Extremely important 61%

Somewhat important 34%

How strongly do you agree or disagree with the following statement: The development of solar power and other renewable energy sources, including the financial support needed, should be a major priority of the federal government.

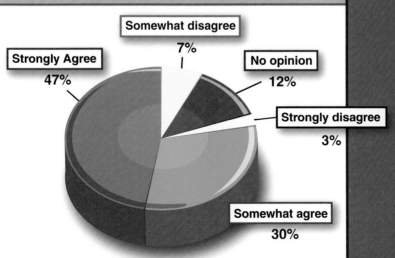

Somewhat disagree 7%

Strongly Agree 47%

No opinion 12%

Strongly disagree 3%

Somewhat agree 30%

Note: Totals equal less than 100 percent due to rounding.

Source: SCHOTT North America, "SCHOTT Solar Barometer Survey," June 2008. www.gosolarcalifornia.ca.gov.

Chronology

1905
Albert Einstein publishes a paper on the photoelectric effect and is later awarded the Nobel Prize for his work.

1891
American inventor Clarence Kemp patents the first solar water heater.

1767
Swiss scientist Horace de Saussure builds the world's first solar collector, which he calls a "hot box."

1861
French mathematician August Mouchet builds the world's first solar-powered steam engine.

1958
Scientist Charles David Keeling sets up a sampling station at the Mauna Loa Observatory on the island of Hawaii and begins his groundbreaking CO_2 research.

1700   1800   1900

1839
French scientist Alexandre Edmond Becquerel discovers the photovoltaic effect, which is the physics behind solar cells.

1968
American scientist Peter Glaser becomes the first to introduce the concept of space-based solar power stations.

1876
William Grylls Adams and Richard Evans Day discover that by focusing sunlight on selenium electricity is produced, which proves that a solid material can convert light into electricity without heat or moving parts.

1980
Paul MacCready builds the first solar-powered aircraft, the *Solar Challenger*, and flies it from France to England.

2007
Photovoltaic manufacturing grows by 74 percent in the United States, and PV installations increase by 45 percent, both of which are among the highest growth rates in the world.

1990
The solar-powered Hubble Space Telescope is launched and begins orbiting Earth.

1991
The Solar Energy Research Institute is renamed the National Renewable Energy Laboratory.

2000
Astronauts at the International Space Station begin installing solar panels on what will eventually become the largest solar power array in space.

1990

2000

2010

1999
Spectrolab, Inc. and the National Renewable Energy Laboratory develop a photovoltaic solar cell that converts more than 32 percent of the sunlight that is focused on it into electricity.

2004
California Governor Arnold Schwarzenegger proposes the Solar Roofs Initiative, which calls for 1 million solar roofs to be installed throughout the state by the year 2017.

2001
Home Depot begins selling residential solar power systems in three of its stores in San Diego, California.

2008
Scientists at the National Renewable Energy Laboratory set a world record in solar cell efficiency with a photovoltaic device that converts nearly 50 percent of the light that hits it into electricity.

Key People and Advocacy Groups

American Council for an Energy-Efficient Economy: The council is an organization dedicated to advancing energy efficiency in order to promote economic prosperity, energy security, and environmental protection.

American Solar Energy Society (ASES): The ASES is dedicated to increasing the use of solar energy, energy efficiency, and other sustainable technologies in the United States.

Isaac Asimov: A world-famous science-fiction author and scientist, Asimov advocated capturing solar power from space-based platforms.

Albert Einstein: A famed German-born physicist, Einstein published a paper on the photoelectric effect in 1905 and was awarded the Nobel Prize for his work in 1921.

Peter Glaser: An American scientist, Glaser was the first to introduce the concept of space-based solar power stations in 1968.

Al Gore: A former U.S. vice president, Gore is an outspoken environmentalist and advocate for eliminating fossil fuel emissions.

Charles David Keeling: An American scientist, Keeling revolutionized the study of CO_2's relationship to global warming by taking measurements on Mauna Loa on the island of Hawaii.

August Mouchet: A French mathematician, Mouchet built the world's first solar-powered steam engine in 1861.

National Renewable Energy Laboratory (NREL): The NREL is the United States' primary laboratory for renewable energy and energy efficiency research and development.

Horace de Saussure: A Swiss scientist, Saussure built the world's first solar collector, which he called a "hot box," in 1767.

Solar Energy Industries Association: A national trade association for the solar energy industry.

Related Organizations

Alliance to Save Energy (ASE)

1850 M St. NW, Suite 600

Washington, DC 20036

phone: (202) 857-0666 • fax: (202) 331-9588

e-mail: info@ase.org • Web site: www.ase.org

The ASE promotes energy efficiency worldwide to achieve a healthier economy, a cleaner environment, and greater energy security. Its Web site offers news articles, a collection of facts, the *eFFICIENCY* newsletter, and a topics section.

American Council for an Energy-Efficient Economy (ACEEE)

529 14th St. NW, Suite 600

Washington, DC 20045-1000

phone: (202) 507-4000 • fax: (202) 429-2248

e-mail: info@aceee.org • Web site: www.aceee.org

The ACEEE is dedicated to advancing energy efficiency in order to promote economic prosperity, energy security, and environmental protection. Featured on its Web site are news releases, ACEEE program areas, an extensive list of links to related sites, and several publications available for purchase.

American Council on Renewable Energy (ACORE)

1600 K St. NW, Suite 700

Washington, DC 20006

phone: (202) 393-0001 • fax: (202) 393.0606

e-mail: info@acore.org • Web site: www.acore.org

ACORE works to bring all forms of renewable energy into the mainstream of America's economy and lifestyle. Its Web site offers news releases, policy descriptions, reports, and links to articles.

American Solar Energy Society (ASES)

2400 Central Ave., Suite A

Boulder, CO 80301

phone: (303) 443-3130 • fax (303) 443-3212

e-mail: ases@ases.org • Web site: www.ases.org

ASES is dedicated to increasing the use of solar energy, energy efficiency, and other sustainable technologies in the United States. Its Web site features news articles, news releases, and reports, as well as a link to the Solar Today blog.

Energy Efficiency and Renewable Energy (EERE)

Mail Stop EE-1

Department of Energy

Washington, DC 20585

phone: (877) 337-3463

Web site: www.eere.energy.gov

An agency of the U.S. Department of Energy, the EERE seeks to enhance energy efficiency and productivity; bring clean, reliable, and affordable energy technologies to the marketplace; and make a positive difference in Americans' lives by enhancing their energy choices and quality of life. Available on its Web site are speeches, congressional testimonies, news articles, news releases, and a search engine that retrieves numerous publications related to solar power.

Interstate Renewable Energy Council (IREC)

PO Box 1156

Latham, NY 12110-1156

phone: (518) 458-6059

e-mail: info@irecusa.org • Web site: www.irecusa.org

The IREC's mission is to accelerate the sustainable use of renewable energy sources and technologies in and through state and local government and community activities. Its Web site offers three different newsletters, a "Resources" section with various publications, and links to related sites.

Massachusetts Technology Collaborative (MTC)

75 North Dr.

Westborough, MA 01581

phone: (508) 870-0312 • fax: (508) 898-2275

e-mail: mtc@masstech.org • Web site: www.masstech.org

MTC is Massachusetts' development agency for renewable energy and the innovation economy. Its Web site links to a special section called *The Power of the Sun*, which is a comprehensive guide to solar energy. Also available on the site are news releases and other publications.

National Association of Energy Service Companies (NAESCO)

1615 M St. NW, Suite 900

Washington, DC 20036

phone: (202) 822-0950 • fax: (202) 822-0955

e-mail: info@naesco.org • Web site: www.naesco.org

NAESCO's mission is to promote efficiency as the first priority in a portfolio of economic and environmentally sustainable energy resources, as well as encourage customers and public officials to consider energy efficiency when they are making energy choices. Its Web site features news articles, news releases, a newsletter, and a "Resources" section that has various publications.

National Energy Education Development (NEED) Project

8408 Kao Circle

Manassas, VA 20110

phone: (703) 257-1117 • fax: (703) 257-0037

e-mail: info@need.org • Web site: www.need.org

The NEED Project seeks to make energy education a priority in schools and colleges throughout the United States. Its Web site features a number of NEED Energy InfoBooks for students of all ages, energy polls, *Energy Exchange* and *Career Currents* newsletters, a searchable energy bibliography, and news releases.

National Renewable Energy Laboratory (NREL)

1617 Cole Blvd.

Golden, CO 80401-3393

phone: (303) 275-3000

e-mail: info@nrel.gov • Web site: www.nrel.gov

The NREL is the United States' primary laboratory for renewable energy and energy efficiency research and development. Its Web site offers a wide variety of information about solar and other types of renewable energy, including a "Student Resources on Renewable Energy" section.

Solar Energy Industries Association (SEIA)

805 15th St. NW, Suite 510

Washington, DC 20005

phone: (202) 682-0556

e-mail: info@seia.org • Web site: www.seia.org

The SEIA is a national trade association for the solar energy industry. Its Web site offers research, news releases, a Solar in the News articles section, industry data, and a solar energy report.

Solar Energy International (SEI)

76 S. Second St.

Carbondale, CO 81623

phone: (970) 963-8855 • fax: (970) 963-8866

e-mail: sei@solarenergy.org • Web site: www.solarenergy.org

SEI's mission is to provide education and technical assistance so others will be empowered to use renewable energy technologies. Available on its Web site are a monthly newsletter, articles about renewable energy, annual reports, and energy facts.

For Further Research

Books

Godfrey Boyle, ed., *Renewable Energy*. New York: Oxford University Press, 2004.

Travis Bradford, *Solar Revolution*. Cambridge, MA: MIT Press, 2006.

David Craddock, *Renewable Energy Made Easy: Free Energy from Solar, Wind, Hydropower, and Other Alternative Energy Sources*. Ocala, FL: Atlantic, 2008.

Alfred W. Crosby, *Children of the Sun: A History of Humanity's Unappeasable Appetite for Energy*. New York: Norton, 2006.

Robert L. Evans, *Fueling Our Future: An Introduction to Sustainable Energy*. Cambridge: Cambridge University Press, 2007.

Rex Ewing, *Power with Nature: Solar and Wind Energy Demystified*. Masonville, CO: PixyJack, 2003.

Stan Gibilisco, *Alternative Energy Demystified*. New York: McGraw Hill, 2007.

Howard C. Hayden, *The Solar Fraud: Why Solar Energy Won't Run the World*. Pueblo West, CO: Vales Lake, 2004.

S.L. Klein, *Power to Change the World: Alternative Energy and the Rise of the Solar City*. Charleston, SC: BookSurge, 2008.

Hermann Scheer, *The Solar Economy: Renewable Energy for a Sustainable Future*. Sterling, VA: Bookscan, 2004.

Periodicals

Patrick Barry, "Greener Green Energy: Today's Solar Cells Give More than They Take," *Science News*, March 1, 2008.

Maria Burke, "Going Solar," *Chemistry and Industry*, April 7, 2008.

Joe Desposito, "Is Solar Energy Really Ready to Rumble?" *Electronic Design*, June 26, 2008.

Mat Dirjish, "Researchers Open Windows of Opportunity for Solar Power," *Electronic Design*, September 25, 2008.

Paul Franson, "First Floating Solar Array?" *Wines & Vines*, June 2008.

Scott Gibson, "Solar Energy: Why It's Better than Ever," *Mother Earth News*, August/September 2008.

James Graff, "Blazing a Trail," *Time International*, December 10, 2007.

William Gray, "We Are Not in Climate Crisis," *ColoradoBiz*, September 2008.

John Gulland and Wendy Milne, "Choosing Renewable Energy," *Mother Earth News*, April/May 2008.

Tom Haywood, "Futurist: Coming Surge in Solar Technology May Make It No. 1," *Natural Gas Week*, April 28, 2008.

Glenn Heglard, "There's Money in Sunlight," *Assembly*, September 2008.

Industrial Heating, "Solar Power Is Hot," August 2008.

Glenn L. McCullough Jr., "Pain at the Pump Now Highlights Need for Energy Secure Future," *Mississippi Business Journal*, September 8, 2008.

Wendy Priesnitz, "Can We Save Money by Going Solar?" *Natural Life*, November/December 2008.

Soll Sussman, "Making a Case for Global Energy," *Latino Leaders*, May 2008.

Breanne Wagner, "Energy Crunch: Army Powers Up for Ambitious Fuel Saving Program," *National Defense*, April 2008.

Internet Sources

CNN, "Interview: Gerhard Knies," August 29, 2007. www.cnn.com/2007/TECH/science/08/23/knies.qa/index.html.

John C. Mankins, "Space-Based Solar Power," *Ad Astra*, Spring 2008. www.nss.org/adastra/AdAstra-SBSP-2008.pdf.

National Renewable Energy Laboratory, "Solar Research," October 25, 2007. www.nrel.gov/solar.

The NEED Project, "Solar," *Secondary Energy Infobook*, 2008. www. need.org/needpdf/infobook_activities/SecInfo/SolarS.pdf.

Solar Energy Industries Association (SEIA) and Prometheus Institute, *US Solar Industry Year in Review 2007*, April 22, 2008. www.seia.org/ galleries/pdf/Year_in_Review_2007_sm.pdf.

Matthew Wald, "Report Calls for Overhaul of Power Grid to Handle Sun and Wind Power," *New York Times*, November 10, 2008. www. nytimes.com/2008/11/10/business/10grid.html.

George Zobrist, "Solar Energy—an Alternative Energy Source," *Today's Engineer*, May 2008. www.todaysengineer.org/2008/May/solar_ener gy.asp.

Ken Zweibel, James Mason, and Vasilis Fthenakis, "A Solar Grand Plan," *Scientific American*, December 16, 2007. www.sciam.com/article.cfm? id=a-solar-grand-plan.

Source Notes

Overview

1. Quoted in Ken Butti and John Perlin, "Horace de Saussure and His Hot Boxes of the 1700's," 1980. http://solarcooking.org.
2. Ken Zweibel, James Mason, and Vasilis Fthenakis, "A Solar Grand Plan," *Scientific American*, December 2007. www.sciam.com.
3. Zweibel, Mason, and Fthenakis, "A Solar Grand Plan."
4. Virginia Lacy, "Generating Electricity from the Sun's Heat," Yahoo! Green, January 22, 2008. http://green.yahoo.com.
5. Solar Energy Industries Association (SEIA) and Prometheus Institute, *US Solar Industry Year in Review 2007*, April 22, 2008. www.seia.org.
6. Quoted in William M. Welch, "Air Force Embraces Solar Power," *USA Today*, April 18, 2007. www.usatoday.com.
7. U.S. Department of Energy, "How Fossil Fuels Were Formed," October 9, 2008. http://fossil.energy.gov.
8. Sven Teske, *Energy [R]evolution*, October 2008. www.greenpeace.org.
9. Union of Concerned Scientists, "Common Sense on Climate Change," November 2004. www.ucsusa.org.
10. The NEED Project, "Solar Energy," *Intermediate Energy Infobook*, p. 23.
11. John C. Mankins, "Space-Based Solar Power," *Ad Astra*, Spring 2008. www.nss.org.

Is Solar Power a Viable Energy Source?

12. Quoted in Port of Entry, "World's Largest Solar Power Plant Being Built in Eastern Germany," June 23, 2008. www.portofentry.com.
13. Quoted in Stephen Koff, "Was Jimmy Carter Right?" *Energy Bulletin*, October 12, 2005. www.energybulletin.net.
14. Travis Bradford, *Solar Revolution*. Cambridge, MA: MIT Press, 2006, p. 98.
15. Bradford, *Solar Revolution*, p. 98.
16. Solar Energy Industries Association and Prometheus Institute, *US Solar Industry Year in Review 2007*.
17. Carolyn Gramling, "Desert Power: A Solar Renaissance," *Geotimes*, April 2008. www.geotimes.org.
18. Lacy, "Generating Electricity from the Sun's Heat."
19. Frederick Reimers, "Oregon Energy," Earth Island Institute, Summer 2006. www.earthisland.org.
20. Bradford, *Solar Revolution*, p. x.

Could Solar Power Replace Fossil Fuels?

21. Richard Somerville, interview with author, December 4, 2008.
22. Quoted in NaturalGas.org, "History," 2004. www.naturalgas.org.
23. U.S. Department of Energy, "A Brief History of Coal Use," October 9, 2008. http://fossil.energy.gov.
24. Martin Hoffert, interview with author, October 11, 2008.
25. Hoffert, interview with author.
26. Hoffert, interview with author.
27. Quoted in Soll Sussman, "Making a Case for Global Energy: Author Robert Bryce," *Latino Leaders*, May 2008, p. 55.
28. Zweibel, Mason, and Fthenakis, "A Solar Grand Plan."
29. Zweibel, Mason, and Fthenakis, "A Solar Grand Plan."

Could Solar Power Help Stop Global Warming?

30. Solar Energy Industries Association and Prometheus Institute, *U.S. Solar Industry Year in Review 2007.*

31. James E. Hansen, Foreword to *Tackling Climate Change in the U.S.*, ed. Charles F. Kutscher. Boulder, CO: American Solar Energy Society, January 2007. http://ases.org.

32. Hansen, Foreword.

33. Quoted in Helen Briggs, "50 Years On: The Keeling Curve Legacy," BBC News, December 2, 2007. http://news.bbc.co.uk.

34. Charles D. Keeling, "A Brief History of Atmospheric Carbon Dioxide Measurements and Their Impact on Thoughts About Environmental Change," *The Winners of the Blue Planet Prize 1993*, p. 73. www.af-info.or.jp/en/blueplanet/doc/list/1993essay-keeling.pdf

35. Michael Griffin, interview, "NASA Chief Questions Urgency of Global Warming," National Public Radio, May 31, 2007. www.npr.org.

36. Richard Somerville, "If I Were President: A Climate Change Speech," American Meteorological Society, August 2008. http://ams.allenpress.com.

37. Zwiebel, Mason, and Fthenakis, "A Solar Grand Plan."

38. Quoted in Port of Entry, "World's Largest Solar Power Plant Being Built in Eastern Germany."

What Is the Future of Solar Power?

39. Al Gore, "Gore's Energy Challenge: 'The Future of Human Civilization Is at Stake,'" AlterNet, July 22, 2008. www.alternet.org.

40. Gore, "Gore's Energy Challenge."

41. Quoted in David Ehrlich, "Solel's New 553 MW SolarThermal Plant," *CleanTech Group News*, July 25, 2007. http://cleantech.com.

42. Quoted in Brian Berger, "Report Urges U.S. to Pursue Space-Based Solar Power," Space.com, October 11, 2007. www.space.com.

43. John C. Mankins, "Energy from Orbit," *Ad Astra*, Spring 2008. www.nss.org.

44. Quoted in TREC, "Press Statement by German Association for the Club of Rome," April 23–24, 2008. www.desertec.org.

45. Quoted in Sandia National Laboratories, "Sandia's Sunshine to Petrol Project Seeks Fuel from Thin Air," December 5, 2007. www.sandia.gov.

46. Bradford, *Solar Revolution*, pp. xii–xiii.

List of Illustrations

Is Solar Power a Viable Energy Source?
Types and Use of Renewable Energy   32
Solar Power Potential: Germany and the United States   33
World Opinions on Energy   34
The Growth of Renewable Energy   35

Could Solar Power Replace Fossil Fuels?
Electricity Generation in the United States   47
Worldwide Fossil Fuel Reserves   48
The World's Top Oil Producers   49

Could Solar Power Help Stop Global Warming?
The Greenhouse Effect and Global Warming   63
Concern About Global Warming Increasing   64
The Steady Rise of CO_2 in the Atmosphere   65

What Is the Future of Solar Power?
Space-Based Solar Power   77
Future Potential of Renewable Energies   78
Americans' Views on Future Energy   79

Index

Ad Astra (magazine), 70
Arrhenius, Svante, 52
Asimov, Isaac, 68, 69
atmosphere, 6
  rise of CO_2 levels in, 52–54, 62, 65
    (chart)

Ball, Timothy, 58
Bellamy, David, 61
biomass, 25–26
  as percent of total renewable energy
    used, 32 (chart)
Bradford, Travis, 22, 27, 58, 72
Bryce, Robert, 36, 40

California Solar Initiative, 32
carbon dioxide (CO_2)
  atmospheric levels linked with fossil
    fuel burning, 52–54
  as greenhouse gas, 17, 18
  recycling into fuels, 71–72
  rise in atmospheric levels of, 62, 65
    (chart)
carbon footprint, 50–51
Carter, Jimmy, 21–22
Club of Rome, 71
coal, 38
  consumption of, 39, 46, 47
concentrating solar power plants
  (thermal energy plants), 10–11
  cost of energy from, 31
  energy cost from, photovoltaic vs.,
    24–25
  land available for, 10
  potential for meeting U.S. energy
    needs, 76
  potential impact on greenhouse gas
    emissions, 62

Damphousse, Paul, 69–70
Daviss, Bennett, 75
Department of Energy, U.S. (DOE), 15,
  59
  on CO_2 emissions from vehicles/

electricity generation, 65
  on photovoltaic potential, 31, 76, 78
Desertec project (Sahara Desert), 71

electricity generation, in U.S., by source,
  47 (chart)
energy
  opinions on sources of, 43 (chart)
  U.S. consumption, sources of, 10
  See also specific sources of
Energy Information Administration
  (EIA), 10
Environmental Protection Agency (EPA),
  26, 65
European Renewable Energy Council, 76

fossil fuels, 7, 15, 36–37
  estimated depletion dates for, 46
  evolution of use of, 37–39
  as finite resource, 39–40
  as percent of total U.S. energy
    consumption, 10
  solar energy vs. energy stored in
    reserves of, 43
  worldwide reserves of, by region, 48
    (chart)
Fthenakis, Vasilis, 10, 31, 36, 41, 42, 55

geothermal energy, as percent of total
  renewable energy used, 32 (chart)
Germany
  available solar energy in, U.S. vs., 33
    (chart)
  installations of photovoltaic cells in, 32
  solar energy efforts in, 20–21, 22–23,
    56
Glaser, Peter, 69
global warming, 63 (illustration)
  attitudes on, 64 (chart)
  debate over extent of problem, 54–55,
    56–57
  potential of solar energy to reduce,
    17–18
Golden, Roger, 29

Gore, Al, 28, 66–67
Gramling, Carolyn, 24
Graps, Amara, 8
greenhouse effect, 51, 63 (illustration)
greenhouse gases, 51
    reduction in, from switch to solar
        thermal power, 62
Greenpeace USA, 45, 76
Griffin, Michael, 54–55
Grose, Simon, 20, 73

Hansen, James E., 51–52
Hayden, Howard C., 28, 66
Hayes, Denis, 22
Herschel, John, 9
Hoffert, Martin, 39, 40, 46, 60, 73
Hughes, Clark, 66
hydroelectric energy, as percent of total
    renewable energy used, 32 (chart)

Industrial Revolution, 17, 38,39,54
Inhofe, James, 60
Intergovernmental Panel on Climate
    Change (IPCC), 54
International Space Station (ISS), 13–14

Japan, solar energy efforts in, 22

Keeling, David, 52–54, 65
Keeling Curve, 52, 53
Kennedy, Robert F., Jr., 75
Knies, Gerhard, 70–71
Kutscher, Charles F., 61
Kurzweil, Ray, 20, 47

Lacy, Virginia, 10–11, 25, 45
Las Vegas Sun (newspaper), 29
Limbaugh, Rush, 30
Locke, Susannah, 29
Lurie, Neal, 59

Mankins, John C., 19, 43
Manning, Andrew, 53
Manning, Lawrence, 68
The Man Who Awoke (Lawrence
    Manning), 68–69
Mason, James, 10, 31, 36, 41, 42, 55
Massachusetts Institute of Technology
    (MIT), 44, 76

Massachusetts Technology Collaborative,
    30
methane, 26
Michaels, Patrick J., 50
Michaels, Robert J., 44
Million Solar Roofs bill (CA), 32
Mills, David R., 40–41
Mints, Paula, 14
Morgan, Robert G., 40–41

National Coal Council, 62
National Energy Education Development
    (NEED) Project, 30
    on biomass, 25–26
    on energy radiating from sun, 6
    on potential of solar power, 18
National Energy Foundation, 39
National Renewable Energy Laboratory
    (NREL), 24, 31
natural gas, 37–38
Newsom, Gavin, 44

Obama, Barack, 6, 74
Office of Energy Efficiency and
    Renewable Energy (Department of
    Energy), 59
oil, major producers of, 49 (map)

PC Magazine, 44
photosynthesis, 13
photovoltaic (PV) technology, 6, 11
    cost of, 24–25, 31
    Department of Energy's goals for, 76
    growth in, 23–24, 31
pollution, from fossil fuels, 46
Port, Otis, 30
Prometheus Institute for Sustainable
    Development, 14, 22, 50

Reagan, Ronald, 21–22
Reimers, Frederick, 26
renewable energy
    American's views on, 79 (chart)
    future potential of, 78 (chart)
    growth of, by country, 35 (chart)
    percent of energy from, by type, 32
        (chart)
    as percent of total U.S. energy
        consumption, 10

Saussure, Horace de, 8–9
Scheer, Hermann, 74
Schwartz, Evan I., 60
Schwarzenegger, Arnold, 74
Sierra Club, 43
Smith, O. Glenn, 74
Solar Electric Generating Stations
   (SEGS), 25
Solar Electric Generating System
   (SEGS), 35
Solar Energy Industries Association
   (SEIA), 14, 45, 50
Solar Energy Research Institute, 21 [ED:
   "Energies" here], 22, 24
solar power
   advantages of, 14
   American's views on, 79 (chart)
   challenges of, 14–15, 44
   methods of capturing, 10–11
   as percent of total electricity generated
      in U.S., 31
   as percent of total renewable energy
      used in U.S., 32 (chart)
   as percent of total U.S. energy
      consumption, 10
   potential of, 76
   resources in U.S. vs. Germany, 33
      (chart)
   space-based, 18–19, 69–70, 77
      (illustration)
   strategies for transition to, 40–42
   viability of, 7, 28–30
Somerville, Richard, 37, 55
Spain, solar energy efforts in, 22, 56
Stechel, Ellen B., 72
Stossel, John, 59
sun, 9–10
   energy radiating from, 8, 28
      in U.S. vs. Germany, 33 (chart)
   historical use of, 8–9
   hydrologic cycle consumes energy
      from, 35
Sunlight to Petrol (S2P) project, 71–72
surveys
   on American's attitudes on global
      warming, 64 (chart)
   on American's views on future energy,
      79 (chart)
   on global support for alternative
      energy sources, 34

Taylor, Jerry, 45, 75
Teske, Sven, 17, 50
thermal energy plants. *See* concentrating
   solar power plants
Trans-Mediterranean Renewable Energy
   Cooperation (TREC), 71
transportation
   recycling CO_2 into fuels for, 71–72
   as source of CO_2 emissions, 62

Union of Concerned Scientists, 17
United States
   available solar energy in, Germany vs.,
      33 (chart)
   electricity generation, by source, 47
      (chart)
   solar power in
      history of, 21–22
      as percent of total electricity
         generated, 31
      as percent of total energy
         consumption, 10
      as percent of total renewable energy
         used, 32 (chart)
US Solar Industry Year in Review 2007
   (Solar Energy Industries Association
   and Prometheus Institute), 14

vehicles
   CO_2 emissions from, 65
   recycling CO_2 into fuels for, 71–72

Waldpolenz solar park (Germany),
   20–21, 56
Wilder, Clint, 59
Willenbacher, Matthias, 21, 56
wind energy, as percent of total
   renewable energy used, 32 (chart)

Zobrist, George, 8, 29
Zweibel, Ken, 10, 31, 36, 41, 42, 55–56,
   74
   on economics of solar power, 74

About the Author

Peggy J. Parks holds a bachelor of science degree from Aquinas College in Grand Rapids, Michigan, where she graduated magna cum laude. She has written more than 75 nonfiction educational books for children and young adults, and has published a cookbook called *Welcome Home: Recipes, Memories, and Traditions from the Heart*. Parks lives in Muskegon, Michigan, a town that she says inspires her writing because of its location on the shores of Lake Michigan.